Packaging Design

Graphics Materials Technology

Packaging Design

Graphics Materials Technology

Steven Sonsino

VNR VAN NOSTRAND REINHOLD
New York

For Jacqueline

Published in the United States of America by
Van Nostrand Reinhold
115 Fifth Avenue
New York, New York 10003

Distributed in Canada by
Nelson Canada
1120 Birchmount Road
Scarborough, Ontario M1K 5G4, Canada

16 15 14 13 12 11 10 9 8 7 6 5 4 3 2 1

Sonsino, Steven
  Packaging design / Steven Sonsino
  176p 29.7 x 21cm
  Includes bibliographical references (p172)
  ISBN 0-442-30303-3
  1 Packaging - Design   1 Title
TS195.4.S66 1990
688.88 - dc20

This book was designed and produced by
John Calmann & King Ltd, London

Designed by Wade Greenwood

Printed in Singapore by Toppan Printing Company Ltd

Contents

Preface

Packaging design is one of the most exciting areas in the world of retailing: multi-disciplinary, multi-industry, it crosses numerous boundaries. All the work that goes into one pack design is aimed at that moment when the buyer sees the pack and recognizes it from any advertising the maker carries out, or is compelled to buy it because of the power of the pack design alone.

Despite the wide-ranging, accepted forms and functions of packaging, the business is still very subjective. Much market research needs to be carried out to test even the best designer's thoughts. No designers should foist a design on the public without a basis of objective data. This is because no two consumers are alike. Everyone behaves differently and expects different things from what they buy. In addition, consumers buy from a range of stores, which vary in how they value the different aspects of the goods they retail, from price, in discount stores, to quality in high-class department stores.

Another aspect of the book that bears early appreciation is the apparent concentration on food and beverage packaging. This is because food and drink packaging represents 90 per cent of the packaging market and most of the exciting developments in the industry over its relatively short lifespan have been in packs for food and drink. These ideas then percolate through to other forms of packaging.

Packaging design offers the talented designer many opportunities. First there is the potential of using the different packaging materials and techniques of packaging. There seems to be, in colleges around the world, an overall concentration on the graphic aspects of design, drawing, making models, typography. While this is important, packaging design offers many more creative opportunities than that. Printing the same design on a variety of plastics films or using different printing techniques to reproduce those designs can bring wholly new effects and images to the buying public. And only through investigating those materials and methods with packaging specialists can the designer discover them. These aspects of technology and technique are all too easily forgotten by new designers straight from college, even if they have specialized in pack design.

The importance of packaging design to a product is shown by two major product examples: the Coca Cola can and the squeezy tomato ketchup bottle. Both are successful, high-volume products and yet the makers have considered it necessary to change the packaging design often and to great effect. Their success is no accident though. It came about through solid research into the marketplace, a great deal of timely advertising, and an understanding and use of the latest technical developments in can-making and filling. Not all manufacturers would be willing to tread this path, and this can result in a stagnation of design, with symbols and motifs, materials and methods staying the same, year in and year out.

A further observation is that the US and Western Europe, particularly the UK, are leading the world as the multi-disciplinary packaging designers of the moment. Of course, there are notable exceptions. Japanese designers are very good at coming up with new technological feats, such as the can for sake that cools the drink when a special ring tab is pulled or pressed. And German designers seem to have a very good understanding of the technical processes involved in packaging and so account for them in their design work. Within the German packaging industry itself are some very strong manufacturers of packaging machinery, such as Bosch and Windmoller and Holscher, so maybe this is not so surprising.

French designers in recent years have excelled in packaging of fresh foodstuffs. Packs and techniques for packing and keeping fresh French salads have been very successful there for some years now. So perhaps in the not too distant future, especially with the Single European Market of 1992 opening up the possibilities for greater distribution of goods, we shall see an increased role for packaging design around the world.

The themes of how to brand products across national borders and the problems of product distribution in the next century should be high on the curriculum for the pack design students of today.

Acknowledgements
A project such as this could not have been undertaken without the help and work of many people. First, I would like to thank Pauline Covell of the UK's Institute of Packaging, who introduced me to the world of packaging design. I should also like to thank my editor at John Calmann and King, Elizabeth Thussu, and the innovative work of the design team of Michael Wade and Nicholas Greenwood. I am also deeply indebted to the many packaging design and manufacturing companies from the UK, US and Europe who provided us with the best examples of their work to include in the book, and to Jeremy Horner for his photography. In addition, grateful acknowledgement is made to Robert Opie for providing the photographs on pages 142-3 and to Wolff Olins for those on pages 8, 11, 12 and 41.

Introduction

The most basic function of packaging is to preserve and protect the product it contains, but in a changing society, packaging is increasingly called upon to fill a more complex role. Packaging now has to function as part of today's highly competitive marketing and retailing world. In today's self-service environment, packaging has to sell the goods it contains. After market research, promotion and distribution, the product arrives on the shelf. The final step, from shelf to shopping basket, is significantly affected by the product's packaging.

The enormous variety of materials open to the modern designer - from traditional glass and cartons, to the more recent developments including polystyrene tubs and plastics films - makes packaging design one of the most exciting aspects of the retail world, even in the own-label sector as shown here in a range of products from the UK's Tesco and Marks and Spencer stores.

The changing face of retailing

Selling to the consumer is no longer a personal interaction over a counter. Before the super-market, foods would be weighed individually from bulk and wrapped in functional, plain brown paper. However, most of the products to be found in the supermarket are pre-packaged. Self-service has meant a transfer of the role of informing the customer from the the shop assistant to the advertisement and the pack itself. The pressures of competition mean that, more than ever before, product pack-aging now has great importance. To attract the customer, it must convey a unique sales message, promoting the product it contains, but at the same time it must remain informative. The fight to be first into the shopping basket has become a full-scale market-ing war, for although shoppers' time has decreased, the number of product lines on the shelf has increased ten-fold.

Given the overwhelming choice of products in the typical supermar-ket and the lack of time to make purchases, the pack must instant-ly communicate what it contains, using appropriate, easy-to-read text and simple images. More than this, it must convey leadership within the sector and fit comfort-ably within the manufacturer's overall image. The pack must also tie in with any other marketing activity - television or magazine advertising, for instance. If a cosmetics manufacturer promotes a high quality image for a perfume product, a high quality pack is essential.

Superstore and hypermarket shopping in every sector - from electrical and white goods to do-it-yourself and food and drink categories - has given consumers a tremendous choice of products and equipment. The trend towards a better quality of life in the Western world puts ever more emphasis on presentation and value for money. According to US agency Mittleman/Robinson Design Associates: "Packaging is crucial. It's the silent salesman. It's the last thing the customers see before they make a purchase decision." Another US design consultancy, Lister Butler, confirms this, adding that the package is a product's alter ego. "Without its package, a product does not exist."

The sheer volume of products on the store shelf means that packaging designs have to work harder than ever before. One of the most competitive areas for the designer is the hair care market, where big name brands compete against own-label goods, as seen here in the ASDA store at Watford, UK. (Photo Fitch RS)

Packaging and social change

In many Western families now both parents work and the weekly visit to the shops has become the monthly trip. In France, the rise of huge hypermarkets such as Carrefour is characteristic of this trend. In addition, freezers and microwave ovens have made their way into more homes, bringing their own demands on packaging.

Changes in society also have their influence: the size of the average family has altered, creating a trend towards single-serve portions, while lack of time has led to the complete ready-meal. The availability of plastics for packaging has enabled the production of moulded trays for TV dinners.

The increasing age of the Western world's population, the growing numbers of people moving away from home at an earlier age and the high divorce rate have all contributed to the smaller family unit. So single-serve and family packs now compete to share the same, valuable shelf space, and size and shape matter more than ever to the pack designer.

Changes in the structure of the family unit has meant that time to prepare foods has been lost. Spurred by the sales of microwave ovens and freezers this has lead to a greater demand for portion-packed and ready meals, such as these from Tesco, UK (designed by David Davies Associates) and the US Mrs Paul's label (designed by Lister Butler).

Consumer characteristics

Studies have shown the increasing dominance of three types of consumer. An increasing majority are subsistence buyers, whose purchasing patterns and attitudes reflect their relative poverty. Their numbers include a large number of the old and the unemployed. Their purchases are almost exclusively the necessities of life, such as food, clothing and housing. The major consideration for these people is cheapness, rather than quality. Packaging has an important part to play in portraying low price. Graphics, banners and promotional piggyback packs bearing money-off messages can almost effortlessly sell goods.

The second consumer group comprises discriminators, who rank quality above all else, even in times of recession. To discriminators, the organization selling the goods is as important as the products themselves. Packaging has no direct control over this, but it can influence the retailer who sells the goods. The UK store Harrods or the US Bloomingdale's, for instance, which are generally regarded as purveyors of fine goods to the well-off, may reject a product line if the packaging does not reflect their high quality image.

Other important factors to discriminators include the social responsibility of the retailer. A growing number of consumers will not buy goods from South Africa, for instance. And a large number of buyers are prepared to pay more for a product if it is ecologically sound. The concern over CFC propellants in aerosols in the late 1980s is a potent example. (This is discussed in more detail in Chapter 3.4)

In this category, also, the glass bottle and jar markets have been helped enormously by the introduction of recycling projects such as bottle bank schemes in the UK and Italy. In Italy, banks are bell-shaped and so are called "campagnello". In Scandinavia, and lately in the UK, machines have been introduced into shopping centers for the reclamation of tinplate beverage cans. Consumers simply place their used cans into the mouth of the machine. Now plastics bottles can also be recycled. Politicians are only too eager to claim the green vote - in West Germany, for example, the Green Party is represented in parliament - so conservation and recycling have become very important issues.

Some consumers are more influenced by the status of the retailer than they are by the products themselves. Buyers at Bloomingdale's in the US and Harrods in the UK, for example, demand that packaging comes up to very high standards to reflect that desired status. Here, corporate identity has been carried through in a range of items.

It is not just discriminators, however, who worry about the ecological aspect of packaging. General public concern is growing: fears about safety and the environment are putting pressure on product manufacturers to limit their output or to design products using recyclable or biodegradable materials only.

The final group of consumers, known as hedonists, tend to ignore economic recession altogether. The group includes people from widely differing areas of society, who spend large sums on credit and largely ignore marketing trends, buying simply what they want at the time. Because it would be far too difficult to predict how these buyers might respond to packaging or advertising, designers should consider this group much as politicians regard floating voters.

Packaging design today

Packaging has become one of the most exciting and challenging areas in the design world, with the fast-moving pace of development in both graphic design and materials technology, which is continually creating new possibilities in shape design and use of materials. As in the past, the need for imaginative pack design - the use of colour, typography, images and logos - remains at the heart of the design profession. Now, however, modern packaging designers must also keep up with the potential - and constraints - of the many new processes and materials being developed. In the USA, for example, a recent watershed was the development of the squeezy plastics tomato ketchup bottle, which resists the natural acids in the sauce and prevents the ingress of that most-damaging of gases, oxygen. In addition, there is a host of new laminated and metallized films for the packaging of dry-roasted snack foods, among other things, and special printing processes and inks that complement the new materials.

The innovative pack designer can now choose from a vast range of materials - from the traditional board - and glass-based products, to plastics, and combinations of plastics, paper and foil. The best results are always startling, never dull.

However, the best material for a product's image is not always the best for protection and the whole world of pack design is based on a compromise between these two factors. In other words, there is always a trade-off between the practical considerations of packaging and what is or is not possible on the graphics side.

Information

The need for packaging to inform has become a significant part of the design. Nutritional labelling is only one instance of the complex legal requirements a designer must handle. The ubiquitous barcode is another. However, such constraints can be used creatively. One designer recently challenged the perception of the black and white striped barcode as a problem: he converted the thin stripes of the code into a bed of reeds, complete with duck and water lilies, and incorporated it into the overall design of a whisky label.

For the benefit of retailers and distributors, packaging should give certain standard details, such as the consignment to which it belongs, its destination, and details required by the laws of the target country. And finally, to satisfy all users, it must also protect its contents effectively and explain how to use them. So the designers of a pack must now keep an eye on its entire life-cycle - from promotional concept to the practicalities of transportation, containment, storage and display, as well as the details of eventual end use.

One of the most significant developments of the past few years has been the high-barrier plastics bottle, from companies such as American Can and CMB, formerly Metal Box. This enables sensitive yet aggressive products such as ketchups, mustards and jams to be packed in convenient, squeezable containers, often with a shelf life of up to a year or so. Here, the successful Heinz brand identity is echoed in the own-brand version.

New techniques

The trend towards healthy eating and living in the West has given packaging a chance in a battle believed lost: fresh produce versus processed foods. The concern over additives, or E-numbers, has made processed foods seem less attractive, while new packaging techniques, such as modified and controlled atmosphere packaging, can help to prolong the shelf-life of fresh produce.

The packaging of meat, fish, fresh pasta and fruit and vegetables could not now be simpler. The goods are simply placed in a deep polyvinyl chloride (PVC) tub, a mixture of gases including oxygen and carbon dioxide is blown across the top and the pack is sealed with a permeable PVC film to allow the food to breathe. Strawberries can be kept for a week to ten days, as opposed to the three or four days they would last without the process. Initially, designers could not do very much with these totally transparent packs, but now major companies are decorating them by printing between the layers of the PVC in the sealing lid. However, there may now be evidence that this form of packaging fresh food can encourage the development of listeria, a serious form of food poisoning.

Food irradiation, too, is a new development in food preservation, that is, however, subject to the consumer's general concern about radiation. At present the harmonization of the UK with most of Europe and the USA over this issue looks a little way off. The irradiation process, carried out after packaging has been completed, can also affect packaging. Glass, for example, sometimes becomes discoloured and dark patches form during the treatment.

Packaging design should often extend to the product itself in some shape or form, especially on products such as this Skyway soap tablet that will be used every day once the outer carton has been thrown away. This disposes the consumer to remember the unique moulded skyscraper design (designed by David Davies Associates) and so to buy the product again when prompted by the carton in the store.

Future developments

Whatever the process, new or old, and whatever the pack, today's shopper demands fresher and more convenient food products and stronger, yet lighter packaging. This requires either the development of new plastics, or the improvement of those materials already on the market. Pack designers must be aware of all such developments and changes.

New containers in plastics, designed and developed over the past few years, have changed the shape of conventional packaging. The blow-moulded bottle from polyethylene terephthalate (PET) and now the PET drinks can, jar and aerosol are significant examples. The bag-in-box for liquids is another; initially used only for wines, it now packs fruit juices, medical liquids and even blood and serum.

As far as shape is concerned, plastics are extremely versatile, if expensive for some applications. Plastics can be moulded in a variety of ways to suit an almost infinite number of designs. But an understanding of material strength, and just how far you can take plastics, certainly helps in the design process. Even the packaging success story of the past twenty years has limitations.

New plastics materials such as polyvinyl chloride (PVC) and polyethylene terephthalate (PET) have enabled designers using glass and cartons to come up with new ideas, such as the bag-in-box, used at first only for wine. Some of these ideas have been enshrined in the range of freshly squeezed orange juice from Carleton designed by the UK Michael Peters Group.

Glass

Because of the success of plastics, there have been fewer developments in glass packaging in recent years. But with the launch of an international consortium, unimaginatively called International Partners in Glass Research (IPGR), things look set to change. The group includes manufacturing and design expertise from Emhart Machinery Corp of the USA, the UK's Rockware Glass and Weigand Glass of West Germany.

Glass has always been important as a pack for commodity products, but of late it has become increasingly important as a projector of the high-quality image. Glass packaging also makes use of another aspect of shape design: the label. Here shape and position play a crucial role.

Whatever image designers aim for with the traditional graphic tools, there is nearly always a trade-off over what is possible or practical on the technical or printing side. But again, designers should always bear in mind that the best pack to preserve and protect a product might not be the best pack to project it.

Glass - originally a container for commodity liquids and powders - has now become a symbol of high quality. In this range of cosmetics from Penhaligon's, stock bottles have been decorated with ribbons and high-quality labels to emphasize the excellence of the product range. (Designed by the Michael Peters Group.)

New materials

Other new materials, such as the laminated and coextruded plastics and paper films, together with new printing processes, make the design task more exciting. Flexographic printing, for example, for short-run products can now approach the high quality gravure process. Holograms made from plastics films and foils have also made their way on to packs, mostly cartons, at point-of-sale displays.

Walking around supermarkets is a good way to pick up an overview of changing trends in pack design, as well as to test the mettle of the best designers in the world, for packaging is an international business. The major players in the industry are likely to be the multinational corporations and design houses with offices in the US, London and continental Europe, as the barriers between international trade are constantly being broken down.

There is of course plenty of room for the small, flexible design house, too, able to specialize, perhaps, in one of the more daring aspects of design and packaging. The two should be capable of co-existing with each another.

Not long ago, packaging was an afterthought, albeit a necessary one, to stop products being damaged and to keep them fresh and uncontaminated. But today packaging is vital as a sales and marketing tool. It is therefore steadily creeping up the list of priorities for retailers and manufacturers alike.

CASE STUDY AWARDS

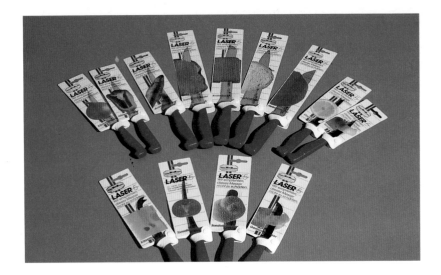

Perhaps the most prestigious awards in the packaging design world are made each year by the World Packaging Organization, with its Worldstar awards. Packs from all over the worlds are shortlisted on technical as well as design criteria and the best packs in consumer and transit packaging are recognized accordingly. Packs from each country are entered in separate preliminary competitions to decide which will go forward onto the Worldstar stage.

In the UK these preliminaries, known as the Starpack awards, are managed by one of the most active design-promotion organizations in Europe, the Institute of Packaging. Entries can be made by the designers themselves, by the pack manufacturers, or by end users, but the packs must be in commercial production. Proofs and prototypes are not eligible.

A stringent series of judging sessions is organized by the IoP, involving independent judges from the design industry as well as from major packaging users such as ICI, Mars Confectionery and Electrolux. These experienced pack designers and specifiers consider very specific aspects of the different packs in the judging process, ranging from the expected protection and preservation characteristics of the packs to sales appeal, graphic design and production quality. Ease of handling and economy of material use are also factors near the top of the judges' checklist.

The 1989 Starpack awards scheme received more than 400 entries from which only two were deemed worthy enough to receive the coveted Gold Star Award. The first of these was won by a pack for a range of stainless steel kitchen knives marketed under the Laser brand name. The high-frequency welded PVC packs from Sheffield-based Thomas Ashton combined what the judging panel called 'real fitness for purpose' with high visual impact. Keeping the handle outside the pack itself enabled customers to test the feel and balance of the knife without being endangered by a poorly fixed blade - additional welding kept the blade firmly in place.

The second Gold Star pack also captured a Technical Innovation award for Rocep Pressure Packs of Scotland, whose convenience container for dispensing a wide range of viscous products offered a highly practical pack. The judges' citation said that the pack offered single-handed, finger-tip control, an adjustable flow-rate and precise application of the product.

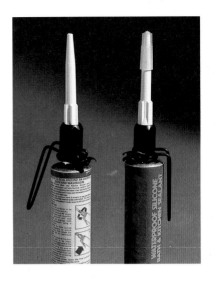

One of five Silver Stars, together with an award from the UK's Plastics and Rubber Institute, was awarded to ICI Garden Products' Grasshopper container. This 5-litre blow-moulded high-density polyethylene jerrican not only stores ICI's latest granular lawn product, it has specially recessed areas in which a dispensing tube and spreader are stored.

A Silver Star was also awarded to another plastics container - Blowmocan's 5-litre pack for Castrol GTX engine oil. The pack has two handles: one for carrying and the second on the side for pouring. This second handle also helps the retailer as it can be used to pick the jerrican from the store shelf. No extra clearance is therefore needed above the container and a better shelf layout can be achieved. In addition, the clear injection moulded stripe down the side helps the user gauge how much oil has been poured into the engine. Full panel self-adhesive labels enhance shelf appeal.

The range of high-frequency welded PVC packs shown opposite, combines real fitness for purpose with high visual impact (Gold Star, 1989).

The Rocep pressure pack, above, allows the consumer to apply high-viscosity products such as sealants, adhesives and mastics without effort, waste or mess (Gold Star, 1989).

A double-neck section in this 5-litre HDPE jerrican allows the air to bleed from the pull-up pouring spout into the hollow top handle, preventing 'glugging' (Silver Star and British Plastics Federation Award, 1989).

The European Flexographic Technical Association award accompanied a Silver Star for the Dixcel toilet tissue range from Metal Closures Venus Packaging. A high degree of printing difficulty was recognized in these reverse-flexo printed, clear medium density polyethylene packs. A line drawing has been replaced with a photographic illustration of two white kittens. Origination costs and the cost of introducing special offer flashes at a later date are approximately 50 per cent lower than normal gravure origination charges (Silver Star and EFTA Award, 1989).
(Photos: Institute of Packaging)

1.1

The packaging designer

Having established that packaging plays an important role in the marketing and selling processes, the question for the designer is: what aspects of packaging influence purchasers? Is it image, quality, or price? And how do packaging designers incorporate the relevant strands of such factors into new designs? All these aspects and many more affect the relationship of designer with eventual consumer. But the job of the designer is made even more difficult by the sheer volume of products that a consumer sees at any retailing location. How can success be achieved?

This range of tinplate paint and primer containers from UK Do-It-All, designed by David Davies Associates, shows how simple graphics, a restricted number of typefaces and the company's rainbow logo can be combined to create a strong and memorable brand image. Branding extends to the design of the retail outlets themselves, as shown opposite, from the shelving to the hanging placards, reinforcing the idea that all the products achieve the same high quality.

The elements of design

There are two key elements to the work of every packaging designer: an ability to understand and work with graphics, and keep up with its constantly changing trends, and an awareness of current packaging materials technology. Most design colleges major on graphics and photography, but there is rarely any detailed study of the different materials and processes available to the pack designer, despite the fact that using the latest materials and printing techniques can give new design work the marketing edge.

Furthermore, designers who understand which materials can and cannot be used in certain situations are the people whose projects will be completed on time and within budget. Designers without that background knowledge may dream up exciting but impractical graphic ideas, causing problems for the production teams that have to create actual packs from the design brief.

The simple transparent pack for this Nishiki Bonsai tree (designed by Brand New), below, shows off the product to the fullest extent possible. Only cellulose acetate and some of the more expensive plastics formulations can be folded without support in this fashion.

For these reasons, when buyers are searching for a good design agency, or when they come to the in-house design department, they need to look for people who possess both graphics skills and a good working knowledge of packaging technology. The reason both elements need to be combined is that much of the work carried out by packaging designers is only indirectly perceived by consumers, who are influenced as much by the colour, shape and texture of the pack as the graphical information on it.

Graphics

On the graphics side there are several basic tools that can be used to influence buyers - mainly shape, typography and colour - and each must be considered in turn and be carefully combined with the others. Shape, for example, has always been an important visual element in graphic arts. It can be used in various ways: there is some evidence that retailers prefer cuboid packs, for example, particularly in the fast-moving consumer goods markets. The ways in which the designer works with the graphical dimensions of the pack is discussed in detail in Chapter 1.4.

Choosing the right material

The designer must first consider the question of what the product is before he or she looks for a material. The designer needs to be informed of the product's physical characteristics and chemical properties:

- is it powder, liquid, paste or solid?
- is it abrasive, fragile, or heavy?
- how big is it?
- does it need protection against water, oxygen, light or heat?
- is it corrosive?
- does it smell?
- does it taint easily - that is, does it need protection against outside contaminants?

A thorough understanding of the product and its protective needs often points directly to the type of packaging material required, especially when the designer has expert assistance from the packaging material manufacturer. Another consideration is the nature of the pack's outer packaging: for example, will it be carton, or tub?

Another example of the use of different materials side by side is this range of bath salts and toiletries designed for the World Wildlife Fund by David Davies Associates. The design leads the consumer to think the preparation is wholesome and natural, and becomes a major selling point in the current health trend.

In the 1960s and 1970s, tetrahedral cartons, such as those shown right, were used for soft drinks and iced confections, but these soon died out because they were crushed so easily. However, a solution appears to have been found for these snacks (designed by Innovations in Packaging).

Marketing considerations

There are also marketing considerations involved in the choice of material. The nature of the competition must be analyzed and the positioning of the product in relation to any similar products. The designer needs to understand what image or message the package should be projecting. Then there is the question of whether the pack should be a bag, a sachet, a pouch, or a simple overwrap and how it will be printed and finished. Most of these marketing considerations should arise from an analysis of who the target customers are, their characteristics and needs.

Production considerations

Concerning the actual production itself, the designer needs to know how many packs or pack blanks should be produced what packaging machinery is available. Is the pack's proposed style compatible with that machinery? What is the maximum production speed and capacity, and does this relate easily to the required annual production being proposed? This essential background can open up dramatic new opportunities for the designer and will prevent many basic errors from creeping into design projects.

The designer must also consider how easy it will be to change the size of pack and whether it is one of a family or range of sizes. Is it likely that promotion packs or trial sizes will be offered in the future? And how will the product be distributed, stored and displayed?

Production and distribution

An understanding of the economics of pack production - and indeed the distribution of packaging throughout the whole packaging chain - can enhance the effectiveness of any designer.

For instance, three-dimensional shape has important side effects. Some packs, such as slim sachets or tubes, cannot stand alone and often a cartonboard tray is used to support and orient the goods face outwards. Soup and cosmetic sachets and other flimsy packs also benefit from a tray support, although these often use ridged and rigid PVC trays instead.

Although these injection-moulded and transparent forms of secondary packaging are usually cheap, they are not normally as effective at promoting the message as printed cartonboard.

Transit packaging

A further point to consider is transportation. Road and rail transport is the usual method of moving goods from manufacturer to wholesaler and retailer, but this tends to treat packaging badly. As wagons are switched at shunting yards, for example, the packs are put through a series of short, sharp emergency stops. At times like these, the question of secondary protective packaging becomes extremely important. The designer has to be certain that, however difficult the journey, when the packs arrive at the other end, the goods will be safe and undamaged enough to be unwrapped and put straight on the shelf.

Methods of protecting goods on trays include stretch and shrinkwrapping using polyethylene or PVC films. The technology of wrapping the stacked pallets has become a science in its own right.

The need for protecting primary packaging with secondary or transit packaging has far more effect on packaging design than most designers imagine. Some pack shapes, even though they may be contained in rectangular corrugated cartons, stack and interlock so badly that far fewer units may be transported on one pallet. This can make transport costs per unit excessively high, even using uniform-sized European and US pallets.

An increasing number of computer programs now exists to help the designer calculate exactly how packs will lock together on a pallet. These so-called palletization programs also give the exact number of packs a pallet will contain. This information can save a great deal of time and money (see pages 38-39)

Finally, and perhaps the guiding factor behind all the designer's work, there is the cost of the total package, including distribution and delivery costs.

The square, waxed cartonboard packaging for the pizzas made by wholefood manufacturer Katie's Kitchen offers an interesting use of a cutaway carton to reveal the fresh pizza itself, recessed into the box. This uses the shape of the carton to emphasise the shape of the pizza. The warm red colouring inside the carton makes the buyer think of hot pizza.

The packaging designer

The skill of the designer comes in welding together the two elements of packaging design - the graphics and the materials - and over the years a number of different kinds of pack designer has emerged. The first type is the small independent consultant, usually inexpensive, almost always local. These designers are usually able to devote a great deal of time to packaging projects as they have low overheads. In many instances, especially in the UK and West Germany, they are more technical than graphics-oriented and have largely come from within the packaging industry.

There are also the large and usually expensive international design agencies, such as Conran Design Group, the Michael Peters Group and Fitch & Co., which approach packaging from the design world. However, they are also able to give increasing consideration to packaging technology. For instance, PA Design, part of the Michael Peters Group, employs a number of engineering-based, draughting designers. The final type, enjoying limited success but perhaps offering the best hope for the future, is the packaging specialist, often drawing staff from both industry and the design world.

How designers work

Although there are different types of designer, they all work in much the same way. Most designers today appreciate the importance of marketing and view packaging more and more as an important aspect of the marketing mix that includes advertising and other forms of promotion. Some consultancies supply only graphics to a client, but they should really be able to provide the technical expertise that will produce the final result. Designers should, in theory, be capable of manufacturing a pack, if need be. Only then can they truly understand the dynamics of the design process.

A customer can find choosing between designers very difficult, even taking quotations and costs into account, so the best usually offer access to two or three satisfied customers for potential new clients to question.

An early idea for a pizza container for Katies Kitchen was to use the shape idea forcefully. Here the Brand New designers effectively took the walls of the carton away to leave the pizza raised on a flat cartonboard base.

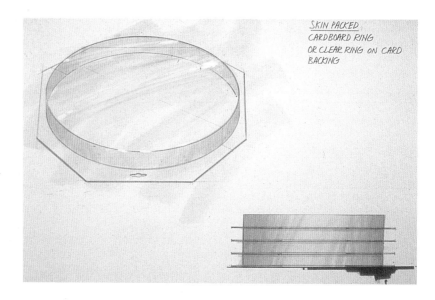

SKIN PACKED
CARDBOARD RING
OR CLEAR RING ON CARD
BACKING

The briefing

When a design consultancy has been chosen, the next stage is the product briefing, where the client tells the designers everything necessary about the product and the market in which it will play. This usually is in the form of a written brief covering the following general areas:

- the product
- any significant market features
- the company's marketing strategy
- the design objectives
- legal requirements
- technical constraints
- the budget
- the deadline for completion

A vital aspect of the briefing is that it should be a two-way process, a conversation between the client and the pack designer, not simply a monologue on behalf of the client. Generally it is much better for members of the design team to know they can put their thoughts and views to the client, rather than be constrained solely by the client's dictates.

The brief should specifically cover the brand's essential characteristics, including colours, typefaces and image. Useful addenda are whether the market has any special predilections or conventions. The high-volume chocolate biscuit market prefers six-packs rather than fivepacks. Cigars, on the other hand usually come in fives with one very notable exception of recent years, Panama, whose "six-appeal" advertisements tied in very well with the pack and, generally, with the image the tobacco house desired.

In the drawing below the cartonboard base was made hexagonal so the product could be hung from one vertex. Both this and the previous design were rejected at the drawing stage, probably on the grounds of storage and display difficulties.

Market research

Having been offered a design project and after completing the briefing sessions, most designers will analyze any existing market research in the early stages and then go out and do their own. The sort of desk research designers often provide includes studying existing published material, the vast majority of which is free - in trade journals and public reference libraries. There are also specific market research studies to be considered, or even commissioned.

These are usually known as Usage and Attitude (U&A) Reports, and they outline how consumers see the client's packaging in comparison with its competitors' packs. On average these U&A studies should be carried out every four or five years, but more efficient companies arrange to conduct them every two years, especially in fast-changing consumer markets.

Designers should visit retailing outlets to check the number of facings a product shows, and to investigate the lighting levels any existing pack receives. It is also necessary to compare the shelf position and number of facings of the competitors' or market leaders' packs, too. In addition, designers should study detailed technical research and existing production methods as well as the existing pack design. They should avoid simply redrawing the pack, with a few cosmetic changes, for the next briefing session.

Evaluation

The new design then needs to be tested on the market. Many proprietary tests come to Europe from the US, but most fail to be widely taken up because they cost a great deal to conduct. There are two basic kinds of test: qualitative, dealing with preference and subjective feelings; and quantitative, where objectively measurable, statistical data is sought, in this case on numbers of potential buyers. In a series of qualitative tests, UK-based agency CWA found that a test pack of British wine featuring a swinging neck label promotion scored highly in in-depth interviews with consumers, when compared with the bottle on its own. As a result, the agency incorporated the neck label into the standard pack design.

Quantitative testing is usually carried out in some central location that can be closely controlled. This offers high efficiency in terms of sampling, as respondents can be recruited from local pedestrian traffic.

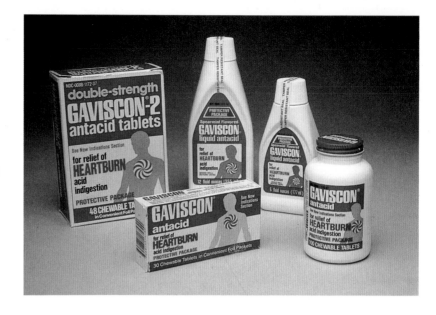

After tests and after the first element of actual designing, it is important to offer all designs for the approval of the client. On some occasions designers seem to show only the best designs, but clients should be allowed to keep their options open for as long as possible. Something that seems undesirable in the early stages could prove invaluable later as other packs are eliminated.

Finishing the project

The designer's job is not over even when the design has finally been approved and - eventually - produced and launched, for the public and trade reactions must be monitored, whether the pack is given a local test launch or a national launch. Some designers believe that when all the basic research has been done, there is no need for further studies. However, designers can monitor whether the product is being properly displayed and keep an eye on the competitors to see if it has provoked a response. Perhaps the best thing to hope for is that the product starts winning shelf space - always a good sign - and attracting attention.

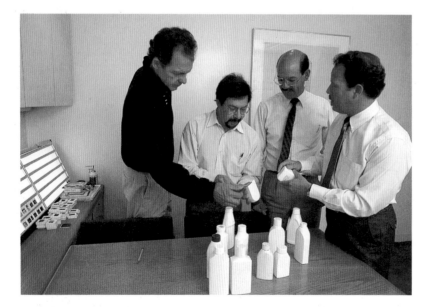

When US agency Gerstman Meyers were commissioned to look at Gaviscon's antacid products, above, they were faced with a reasonably well-selling, but dull range.

After drafting and discarding many new shapes and labels, they decided to keep the range in brilliant white polyethylene plastics bottles with deep blue screw cap closures.

After the launch

As with other creative professions, there are also annual awards for pack designs. The most important UK event, Starpack, is held annually by the Institute of Packaging. There is also an internationally significant label award, made by Fasson, manufacturer of self-adhesive label materials, and there are several other annual events, including one held by chemicals giant Du Pont and another by European label group Finat.

However attractive winning an award might appear for promotional purposes, the best advice to designers is to put them to the back of the mind. Most awards are extremely subjective and, as such, should be no guide for the marketing-oriented pack designer aiming for a more mundane, but important objective: increased sales.

One final point: although some assumptions can be made and intuition used in packaging design, provided that the basic research has been carried out, it is better to proceed on a more rational basis. US-based Peterson & Blyth Associates say: "Forget the assumptions that women are supposed to like soft pastel shades and a container with curves suggests femininity. Colour should be determined by what the competition looks like, not by preconceived notions."

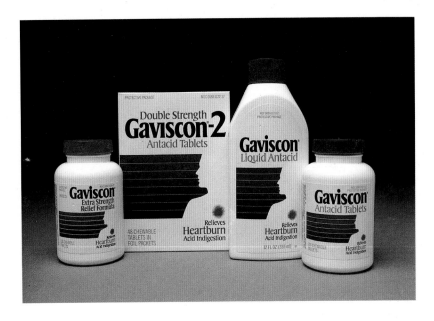

Once the team had chosen the open faced, rather than the recessed style of bottle, the label design came next. The new label, above, was made much larger than the old design and the typeface more open and easy to read. The image of the sufferer is brought to the forefront and enhanced by the graduated blue-black background.

Redesigns

The packaging designer's work consists not only of designing new products for launch but also redesigning established products to attract new markets and maintain market share in the face of competition. Redesigns can range from subtle alterations using established and recognized elements of shape, colour or logo, to a completely new design of an existing product where little is retained other than the name.

The question of how long a pack design should last is almost impossible to answer. Many products are still available today in the same packs in which they have always been available. Or are they? Comparing modern with original packaging for Oxo beef stock cubes, say, it can be seen that although the same base format has been retained, the latest designs are subtly different. The logo has been given a three-dimensional lift by the shadow effect layered behind. The red colour has become blood red, rather than brick red, to help it stand out from the ever-more crowded supermarket shelf. In fact there are a number of subtle changes that ensure the pack is kept up-to-date and noticeable. These - in summary - are the major functions of the re-design.

Because of the speed with which design styles come and go in the pack world, designers must be increasingly aware of the opportunties for redesign. For without prompt attention, products can sink without trace. The redesign, if handled carefully, can ensure not only maintenance of market share, but will often actually increase it if the right note is hit at the right time.

'It looks good enough to eat' could be said to be the ideal to which every food retailer or pack designer should aspire. This saying became the creed of US magazine designer Milton Glaser when he was enrolled by financier Sir James Goldsmith in the task of redesigning the Grand Union supermarket chain. Glaser first tackled one store in New Jersey and from there took on the whole chain, from corporate identity through to own-label packaging.

'The old private label packaging was typical supermarket packaging,' he said 'reduced to a very simple formula: a white can label, usually with mediocre photograph of the food product on it and the old Grand Union logo.'

Now the designers start with a consideration of the nature of the product itself, he affirms. Instead of offering a product that people would normally buy on price alone, Glaser instituted the philosophy that the buyer should feel the product competes with national brandleading products in terms of quality. Clearly this meant a significant change in the pack design philosophy.

The first response was that the products should be redefined within categories. 'Breakfast food would compete against Quaker, the canned goods against Del Monte, and so on,' he said. And so over the past eight years, category by category, Glaser has redesigned over 2000 separate packs, touching everything the retailer markets under own-label.

Redesigning, however, is not a word that Glaser himself uses. 'We haven't had to redesign anything. The underlying assumption we made was that our packaging has to be very clear, very direct. It has to suggest quality without suggesting elitism or fanciness.'

A nod in the direction of the most distinct of 1980s trends, healthy eating, was the yellow band of nutritional information Glaser introduced to every product line. The total line, he adds, was conceived as one proposition, which is why it has interested him for so long. 'You don't do the same thing for apple sauce that you do for paper towels.'

Glaser's background is not, as one might think from his work, straight packaging design. He started in magazine design - founding and later selling *New York* magazine - and some of this editorial vitality he has attempted to bring to his work in the packaging sphere. 'I've been interested in making the packages more editorial, making them look more readable, certainly, but also with a fresher look. Some packages get musty, because they're so serious.'

Where Glaser is most accurate is in his assessment of some of the world's other retailers. To make everything in pack design universal, to signal that the parent store is the manufacturer, is an old-fashioned approach that he argues against. 'We have to dig to find out what it is people want to believe about the product. With apple sauce, for example, there's an old orchard suggesting a made-on-the-farm feeling. That's the reassurance people want.'

In trying to allow his packaging designs to follow the trend towards larger supermarkets, where in-store speciality departments generate even more own-label goods, Glaser attempts to make the packaging look a little less slick than he might otherwise have done. 'You want to look less professionalized, less mechanical. So you don't make the packaging look as though it was overly designed. The idea is to let the product itself do the selling rather than the packaging.' In fact, packaging is secondary in this kind of store, he believes, throwing down the gauntlet: it should be fresh without dominating the product. That is a challenge many other designers and retailers will be forced to follow if Grand Union continues to thrive.

Glaser believes that designers should not follow the same design for peanut butter that you do for paper plates.

Over the past decade, Milton Glaser has redesigned Grand Union's corporate identity - from the outside in.

The hosiery range, left, shows a clean, direct style that many leading brand names have copied.

*1.*2

Computer-aided design

This uniquely shaped glass bottle could have been difficult to design if it had been prepared by traditional graphic means. Its top-heavy shape would have meant preparing numerous models in perspex to check stability and ease of filling. Instead it was designed using a computer as the main drawing tool (right).

With computers as drawing tools, designers are not limited to images they can devise in pen or pencil. They can capture images from photographs, sketches or drawings, and even from video cameras or film. They offer the pack designer the ability to communicate new ideas more exactly, more efficiently and more quickly. They can easily adapt or change complex features - without the traditional drudgery.

Perhaps more importantly, they enable the packaging designer or agency to work directly with the customer. In the past, enquiries for a new product would be handled by several departments before a product designer, using established draughting techniques, produced a single, final specification, which then went to the customer for approval before final production. Using CAD as a marketing tool, the potential customer can become directly involved as the design is created.

Automating the design process

Computer-aided design increases the possibilities of design tremendously, allowing designers at last to be truly creative and come up with new concepts in packaging design, rather than having to worry about some of the more tedious and time-consuming drawing tasks, such as filling in complex backgrounds.

With most computer-based design systems, such as the Quantel Paintbox, designers no longer have to draw by hand all the images they want. Circles and polygons, such as rectangles or squares, may be defined by choosing a centrepoint or corner on the electronic drawing board and simply moving the stylus away from that point. The distance the designer moves the stylus represents the radius of a circle or the diagonal length of a polygon. Because the object is defined mathematically it can be resized and moved with the stylus wherever necessary.

Another problem that no longer faces the computer-aided pack designer is retouching. With such a system, the designer can select a nearby colour with the required intensity by simply pointing to it with the stylus. Then any tool that the system provides - paintbrush, pencil or airbrush - can use that colour.

The Paintbox uses an electronic paint palette and a pressure-sensitive stylus that can create the effect of a paintbrush, chalk, crayon and different kinds of airbrush simply by touching a large graphics tablet, the equivalent of a drawing board. As the designer applies more pressure to the stylus, so the stylus applies more "paint," and the results are displayed on a large computer screen.

Outputting artwork

To get designs out of the computer and on to a pack, transparencies can be taken of the screen and used as artwork, generally at a resolution of about 4000 lines. Increasingly, however, outputting the data on to computer tape or disk is becoming the norm. The next stage - film or platemaking, by output to an electronic scanner that makes colour separations - is relatively easy. Because of the growing number of systems able to achieve this, the cost of electronic reprographics is continually falling. And as the price falls more and more companies are installing computer-aided design systems.

The use of computers to prepare so-called wireframe models enable the modern pack designer to check all the physical characteristics of a bottle, such as capacity and strength, without resorting to the traditional moulding of a perspex model. And even at this early stage in the design process, the computer can show where the label will appear. (Designed by Brand New)

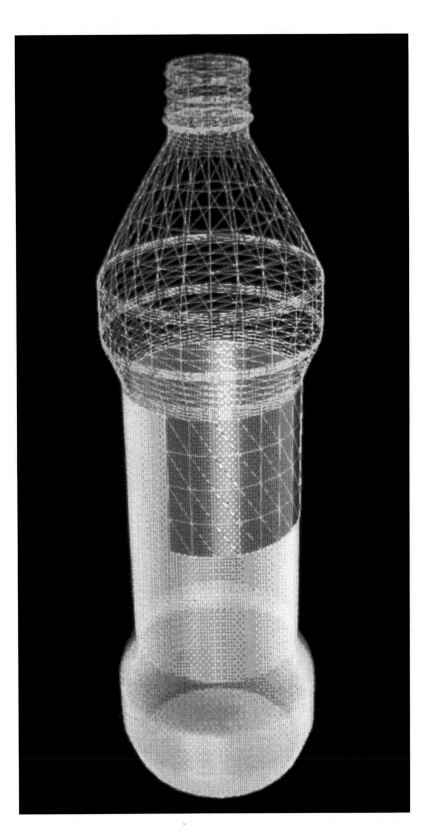

Saving time

One of the first UK-based design agencies to use a CAD system (the Quantel Paintbox) was Towers Noble, which has been able to cut pack design time from weeks to hours, and now to minutes. One cigarette pack redesign for Gallahers took 22 minutes. Towers Noble say that conventional packaging design techniques seem stilted in comparison.

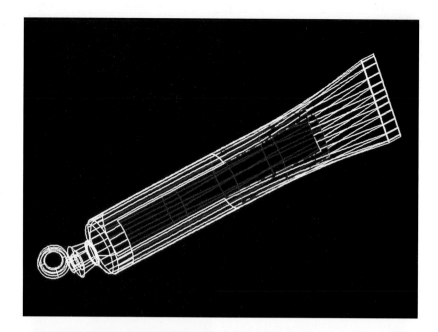

At a seminar on computer-aided packaging design in 1985, design bureau BPCC Video Graphics, in conjunction with specialist pack designer Siebert/Head, completely redesigned a range of German soups for the UK market, again using the Quantel Paintbox. The exercize, using Knorr soups, was completed in less than six hours during which BPCC generated and demonstrated around 30 four- and six-colour images of the soup sachets. These could be made into transparencies or colour separations to pass directly to an electronic scanner. The pack brief was put together before the seminar by Siebert/Head to save time, but the completion of the exercise in a day by the BPCC team shows how much valuable time such systems can save.

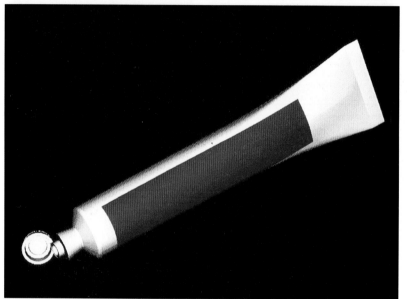

Here CAD is used for designing a product as complex as this collapsible laminated toothpaste tube with integral cap and lid.

The wireframe model - a three-dimensional representation on one of Intergraph's powerful graphics systems - has been overlaid with an outer skin in the second shot.

The power of computers can be appreciated from the third picture in this series, which might be mistaken for a pack shot. The CAD system can rotate the model at will and even place it inside other packs, such as the carton outer. (Designed by Intergraph, UK)

Carton design

Another successful CAD system, "Vista," from the Scitex Corporation, has been used for a number of years by the Reedpack-subsidiary, Field Cartons, to generate graphics for carton design. The designs are composed on-screen using the basic elements of text, line drawings and continuous tone material. As with the Paintbox, these elements may be constructed within the system or borrowed from a library of images. Items can also be captured from photographs, videotape or film.

The system has been used to produce designs and then full-size carton blanks in a few hours. Another advantage is that these systems, notably one from Lasercomb America, reduce and sometimes remove entirely the tedious, time-consuming and potentially error prone handwork associated with producing actual samples, for instance. The computer on which the design is created then controls the laser devices that cut out the carton shapes.

Benefits of CAD

A long term advantage for companies using computers is that computer systems can gradually "learn" which designs are most effective in which systems when they are programmed by top designers. When these designers leave, the company suffers no setback through loss of expertise because the computer has stored a great part of that knowledge.

Because of the fast, interactive nature of these computer systems, the client can be present at the design stage to approve progress or suggest alterations. This highlights another function of computer-aided pack design, as a customer relations tool, impressing on customers the designer's ability to meet their needs quickly and efficiently. This issue of confidence in the designer is increasingly important as the packaging design business becomes ever more competitive. For any designer or independent design group to undertake a range of design projects, it is important to enable clients to feel assured that the project will meet the brief on time and prove successful in the market.

Some critics of CAD systems say that they are not as flexible as the traditional graphic artist. This redesign by US Peterson and Blyth of the Sure and Natural pack shows finely graduated colours, perfectly curved text and an exactly matching scroll that would be very difficult to achieve except by the most experienced artist.

Computer-aided manufacture (CAM)

Some of the first designers to apply CAD were architects and engineers, because their designs required complex drawings and a multitude of associated calculations to be carried out, which exploited the computer's ability to process information very quickly. However, its ability to carry out control processes, especially in manufacturing industry is now being used. For designers, there is now the growing prospect of linking CAD systems to manufacturing plant to directly control the production of packaging, for example, in making container moulds.

In the glass container industry, United Glass, part-owned by US-giant Emhart-Owens, and Rockware Glass have both invested heavily in CAD for many years. Screen-representations of standard elements, such as neck threads and champagne bottoms, can be stored in a database to be called up when necessary by the designer. Using finite-element analysis programs, data concerning the physical performance of the packs can be identified before the bottle is moulded, and the design subtly altered to counter any potential weakness. Even such a small change as moving the centre of gravity of a bottle could have important repercussions for the efficiency of the filling line.

CAD/CAM

The process of linking design directly to the manufacturing process is known within the computer industry as CAD/CAM: computer-aided design/ computer-aided manufacture, combining two types of system - two-dimensional graphics and three-dimensional modelling. The resulting hybrid systems can produce high-resolution, Paintbox-type graphics, but wrapped around packs and objects modelled mathematically in three dimensions. The two images can be separated at will and colour separations produced of the label, while the modelling information is used to control, for example, mould cutting for bottles. There are clearly massive production cost savings to be made with CAD/CAM.

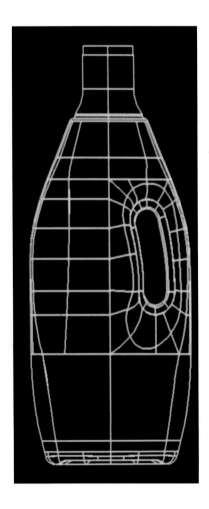

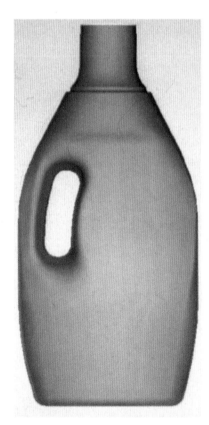

Powerful graphics systems

One of the most powerful of the latest generation CAD systems, used for cartons and glass containers among other things, is Aesthedes from Claessens Product Consultants of the Netherlands. The system, whose UK users include Marks and Spencer and design agency Holmes and Marchant, seems bewildering at first. It has three graphics screens, three smaller text screens and offers about 560 buttons, one for every artistic tool a designer might need, from pencil to airbrush. The basic design can be rotated and seen from all sides, and offers the chance to see what new designs might look like next to competitors' products on the shelf. The Aesthedes can output to Scitex plotters as well as producing second generation originals for scanning and printing. Like the Quantel Paintbox, it can output 16 million colours.

Claessens Product Consultants initially developed the system for its own designers - the company now has at least ten machines for its own use. Its design clients include Heineken, for which Claessens produces packaging worldwide, and James Burroughs, with its Beefeater Gin.

Plastics materials have given designers the ability to mould even the most intricate of shapes. This fairly simple container for a clothes conditioner, left, designed using the Deltacam system, contains a cut-out handle - how does that affect capacity, ease of filling or stability? And how will it look on the shelf? By building up the sequence from wireframe to smooth wireframe and full 3D elevation, the designer can test many thousands of containers in hours rather than months.

The future of CAD

Until recently the major excuse for not adopting CAD for graphic design was expense and the fact that it was not possible to follow the design through electronically to production without sophisticated equipment and a very large bank balance. However, the cost of computer hardware is tumbling dramatically, and CAD systems will soon cost less than a designer's basic drawing equipment. So besides enabling the designer to be more creative, these systems will also prove themselves cost-effective partners in the design studio.

With a limited number of exceptions, few design consultancies have investigated the real possibilities of computer-aided packaging design for the 1990s and beyond. While it may not always be worthwhile at present to develop designs on computer systems, further applications will be studied ever more closely by specialist programmers and hardware designers and eventually most packaging designers could be working on the computer screen.

The Aesthedes system, showing the three-screen workstation.

The introduction of the computer into the design studio has changed the way the modern designer must work, but has not altered the basic design skills he or she needs. Because the computer is nothing more than a tool it must be made to work the way that humans do. It should store and use information that all designers are familiar with and can manage.

The program first designed by Rockware Glass in 1977 for its own use was structured so that the designer was asked a series of questions. These questions, step by step, enabled the designer to build up a picture of the product he or she had in mind. This initial system was very mathematical and the screen displays were simple two-dimensional representations of product designs.

The essential data the computer incorporated into each design included the following information:

● the customer's name to ensure copyright;
● a container description and product type to make very clear the intended use of the container;
● the glass colour, which determines the density of the glass;
● the type of closure to be used (around 400 caps and closures were
● stored on the system's database;
● the sidewall and base thickness of the container;
● the capacity of the container and its fill level - often referred to as the headspace - to allow for the expansion of the contents under high temperatures;
● and finally the type of surface stippling required, if any.

Then the designers chose a selection of the basic designs for the glass finish - that part of the container that supports the closure - and the other shape options. These included the shoulder, the body itself and the base insweep, the radius which the side makes as it becomes the base. Significant dimensions from the design brief, artwork, samples or models were added and then - fully programmed - the computer could calculate the required capacity, bottle weight, the centre of gravity (full or empty) and a number of other functions all within seconds.

Alterations to the basic design could also be made rapidly. The complete design data was stored, mainly on magnetic tape, and in this format could be used to help define a specification for new transit packaging.

Soon the company's design business grew and the self-designed software could not cope with the diverse products its clients wanted new containers for, so Rockware moved to a more powerful system, an Apollo Domain workstation, in 1986. The software the company chose from the engineering world was also more powerful: DUCT (Design Using Computer Technology) and DOGS (Drawing Office Graphics System).

This CAD system enabled Rockware's designers to use colour and move to three-dimensions, offering an instant feedback to the customer on the state of the latest project (see figure 1.) It produced colour images of bottles and jars in three dimensions, working from flat 2D images and wireframe data. The models could be back or side-lit to simulate shelf conditions and the company spent long months working on recreating the translucent and transparent effects of glass on screen. Eventually even this was achieved (see figure 2.)

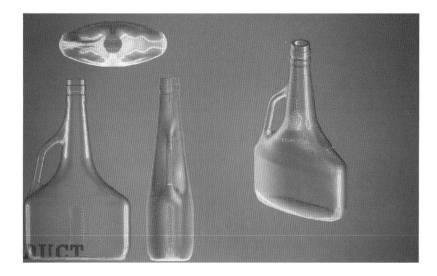

Figure 1

The 3D system cuts production time for a fully visualized bottle, such as the Monterez design shown here, from about five days to a matter of hours. Since it embarked on using computers in the design studio almost fifteen years ago, the company has developed well over 20,000 container specifications. If it were still using traditional design and draughting methods, the company calculates it would still be producing them well into the twenty-first century.

Figure 2

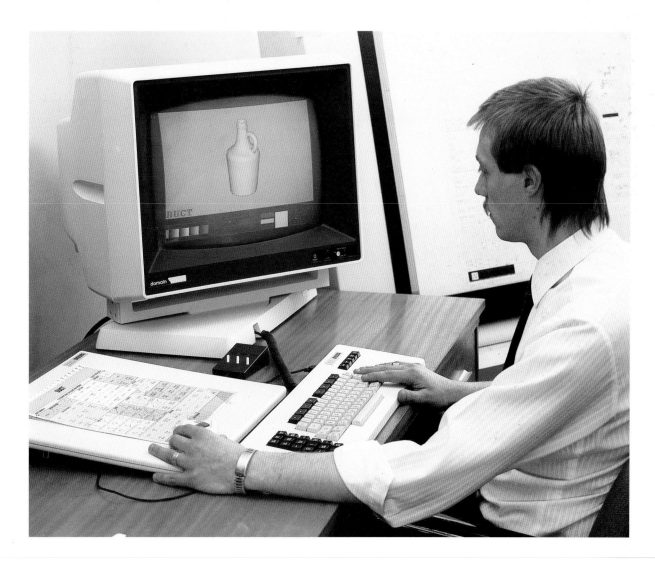

The USA, home of the high-power computer workstation, is also a major user of computer-aided design for packaging. Peterson and Blyth Associates, for instance, use the Contex CAD System - which can store illustrations, photography and logos for easy recall - but not without important restrictions. The designers never allow the system to provide what the company calls 'the design direction', by simply browsing through existing designs and ideas. After all, the machine is only a tool. Once a design direction, or brief, has been defined by the designers and the client, the computer system becomes the basis of a dialogue between them.

Some clients use a remote computer terminal, so that they can view in their office what stage the design team has reached. If they wish, they can make corrections or alterations using a light pen to 'write' comments on a video monitor.

The planogram function enables clients and designers to assess the shelf-impact of a new design against existing products and competitors' ranges.

This colour thermal printer can generate high-quality proofs rapidly. Type is legible even at 6 point.

Reviewing a number of computer-generated models is easy, according to Peterson and Blyth. Instead of needing to generate expensive silk-screen composites, a thermal printer attached to the system can produce a near-Cromalin quality print that simulates the product on the shelf. This planogram can also incorporate a competitor's products, so the product manager can judge how effective the new product will be against existing packs in the supermarket.

Unlike many of the low-resolution CAD systems, which can only be used to generate approval for designs, Contex creates an electronic mechanical, specifying data about the design, which can be passed to a Scitex or Hell scanner, for the generation of colour separations. This time-saving feature is becoming increasingly important for clients who demand an increasingly fast design service. The system can also generate a hardcopy design mechanical, if the client prefers.

CAD systems make it easy for designers to show clients how a pack design will look in its 3D format.

A scanning and plotting workstation can be used to input monochrome line artwork. Logos can be scanned into the system for instance and incorporated into a design, while high-resolution mechanical data can be output on this terminal for approval.

Designing bottles and cartons becomes easy and making changes or trying alternatives is a major advantage over the manual design process, says Peterson & Blyth.

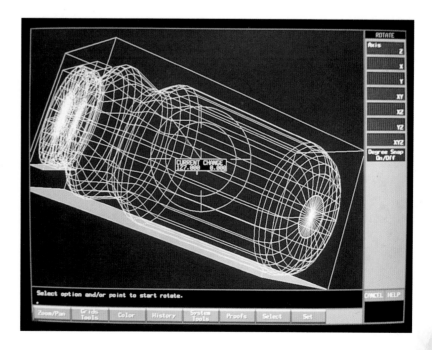

CASE STUDY PALLETIZATION

```
PALLETMANAGER                              SCREEN 03
                                           Collation
PRODUCT CODE : 1234-7
Pack Dimns. :     337 x    137 x 227.9
Pmy. Dimns. :      60 x 130 x   60
Packs/Pallet :    161  Layers :    7 x  23
Cyl. Drawn 20 of 20

                                  REF. NO. :   1
                                  SIDE view of Pack

        337
                                      OPTIONS

                                  PALLET layouts
                                  - press 'RETURN'

                                  ALTERNATE pack
                                        collations
                                  - enter REF. NO.
```

Palletization is the act of fitting as many packs onto a pallet as possible in the most stable configuration for storage or distribution. Until recently this had to be done by guesswork. More often than not, less stable pallet loads were accepted simply because to redistribute the load was considered a waste of time. Now, with the advent of palletization computer programs, things are much simpler.

In action, these computer-aided palletization systems are fairly simple to get to grips with, even for users unfamiliar with computers. Once the dimensions of the various packs have been entered, together with the pallet size, almost any IBM or IBM-compatible microcomputer can calculate the best fit of those containers on the pallet. The screen then displays plan views of the various bottom layers it has calculated and the amount of overhand, or underhang, that the layouts produce. The user can then browse through the available layouts to choose the most suitable, according to criteria including pack size or stability.

```
PALLETMANAGER                                      SCREEN 08
                                                   Packaging
                                      ---- PACK SIZING ----
                              05  Matl. thicknesses on pack length  2
                              06                            Width    2
     STYLE NO.  1            07                            Height   4

01  Description              08  Gap between each PMY (Length)       0
    B flute case with flaps                           (Width )      0
    top and bottom.         09  Gap Constant (Length)               1
                                            (Width )                1
                            10  S/W end seal (Length)               0
    ---- MATERIAL ----                       (Width )               0
02  Thickness        3       11  Tray height                        0
03  Weight (Kg/Sq.M) 0.4     12  Height Allowances - Headspace      0
                                             Each layer pad         0
04  Cost per Sq.M    0.75    13  Type of Case                    0261

          ---- PACK CONSTRAINTS ----
14  Max. tiers in collation   3
15  Maximum pack dimension   500    17  Conveyor working width       0
16  Ht/Base stability         3     18  Conveyor inner radius        0

Sty 1  Sty 2  Sty 3  Sty 4  Sty 5  Sty 6  Sty 7  Sty 8  Sty 9  Sty 10
B flute case with flaps    top and bottom.
```

The design program can also help in this process by displaying a three-dimensional diagram - known as a stacking report - which shows whether the layout provides a sufficient layer interlock, that is, whether or not the load will be stable in motion. If the user thinks a particular layout would not be effective, he or she can move the boxes on the pallet to see what the effect would be on the pallet's stability. The various layouts can then be printed, with additional palletization data to help the users of the printed output.

Not only can these programs examine the loading of existing packs. They can also examine the influence of pack design on costs of packaging, storage and distribution, and it is here that the programs offer the greatest advantage to the pack designer. The main areas of study are the geometric arrangement of primary units within a secondary pack and the construction of the pack itself.

Consider 12 cartons, each 124.5 x 74.9 x 69.9mm and weighing 350g. They are to be packed in a B-flute corrugated case, with flaps at top and bottom. Other useful data are that estimated production volume of the product contained by the cartons is 150,000 a year and it is not essential that the cartons remain upright. On entering these data into the program the system carries out its calculations and displays the results. Between 168 and 192 cartons can be stacked on pallet, depending on the particular geometric arrangement used.

It is important to note, though, that the maximum pallet load may not be the most economic choice when materials and distribution costs are taken into account. The pack designs can, however, be listed in order of total annual cost and a few simple design changes - to the height of the carton, say - can significantly improve the number of packs (of 12 cartons) on each pallet. Design changes actually made to this carton on current prices of distribution and materials would save £7,800 a year over the original design.

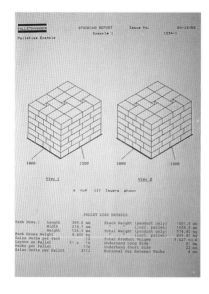

Printed output

Opposite above: collating cylindrical packs or bottles

Left: selecting the outer case style

Right: 3D screen layout

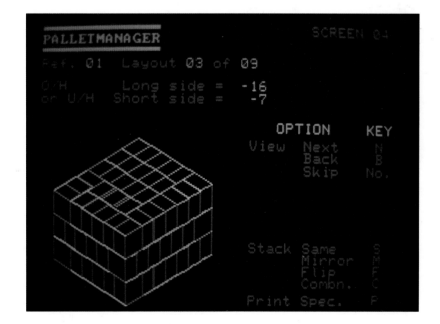

1.3

Packaging and imagery

As part of the marketing process, packaging design is concerned with projecting not only the product itself but also a particular image, whether of a manufacturer's brand or a retailer's own-brand identity. In addition, the product may be targeted to appeal to a particular market segment by evoking certain lifestyle associations. In all this, the packaging designer must work closely with those running the advertising campaign.

Colour and typography can take even the most humble product up-market. This range of quality soups, designed by the Michael Peters Group uses a gold wraparound label to enhance the quality feel of the product. The use of metallic colours almost always has this effect, but it is emphasized in this instance by the fine-stemmed typography.

Packaging and advertising

The impact of pack identity as a reinforcement of television and magazine advertising can be shown by retailer research programmes conducted at the point of sale. A grocery survey carried out by United Biscuits in 1987, for example, had some interesting results and, taken in conjunction with forecasts of budget spending for the future, makes thought-provoking reading about the future for brand versus own-label identity within the pack design forum.

One London advertising agency, Twill Price Court Twivy D'Souza, suggests that packaging is far more important than advertising - despite its clients' initial scepticism. At a seminar on "The serious business of packaging design", the company, which has worked on projects for Brooke Bond, Mates Healthcare and Moulinex, explained that it would much rather take £40,000 from an advert aimed at a prime time news slot to make sure the packaging is better in the long term.

Grocery survey 1987	
Reason for buying	Proportion of respondents (%)
Saw in shop/bought on impulse	42
Recommended by friend/colleague	25
Saw advertised	18
Free sample/trial	9
Other	6
Total	100
Source: United Biscuits	

One of the most famous examples of identity promoted by secondary packaging is the Harrods shopping bag. The very act of buying from stores such as Harrods and Sears makes a personal statement about the consumer.

This attitude makes sense because, after all, the life expectancy or "shelf life" of an advertising campaign is only eight weeks or so, while the shelf life of a pack, in comparison, could be anything up to five years.

Growth forecast in advertising spend

Budget intentions	TV	Press	Packaging design
increasing	11	18	32
stay the same	42	48	44
decreasing	8	9	3

Source: Schlackmans 1988

Target marketing

The marketing of a product is becoming increasingly directed at pre-determined target markets: this is called "segmenting" the market and is usually the result of intensive market research into the characteristics of potential customers. Segmenting can be done by social, economic and geographic factors, the latest and most sophisticated being according to "lifestyle." On the one hand, targeting is becoming more and more specific, pinpointing smaller and smaller segments, but at the same time, especially after 1992 in Europe, markets are becoming increasingly international and the designer has to be careful of being too culture-specific.

Pack design and advertising can also be used to move a product across different market segments, as the need arises. The key points for the designer are when this can be done, how it can be done and with which products. Understanding the target market is then of crucial importance to the packaging designer.

Brand and own-brand identity

With the increasing trend towards high-quality, own-label goods, designers have asked themselves whether there is a future for branded products.

This question was addressed by a group of the UK's top design agencies at a seminar called "The serious business of packaging design" in the late 1980s. The Michael Peters Group suggested that the evolution of own-label suppliers into purveyors of quality goods was crucial in the elevation of packaging design to a serious business; the company suggested that taking own-label goods up-market to compete with branded goods was the only way forward.

Although most of the work in the design industry is still with branded products, these now face strong competition from the own-brand market, as many more retailing organizations are becoming involved with outside design agencies and independent consultants. Revitalizing the image of brand packaging, says the Michael Peters Group, is not simply a question of revolution, or a complete change of identity. The idea is more to refresh the image, to enhance those aspects of a product that put it head and shoulders above its neighbour. The emphasis on brand happens even with commodity products such as beans and cream crackers: consumers are not asked simply to buy baked beans or cream crackers, but Heinz Baked Beans and Jacobs Cream Crackers. Nowhere is packaging more closely linked to advertising than in the mineral water market, where the shape and colour of the Perrier bottle are so strongly tied, in the consumer's mind, to the product that posters and TV advertisements can effectively advertise the product without even mentioning the company name or showing a label.

There is a feeling among design agencies at the moment that shape is very often the greatest competitive edge a product can have. It identifies products on the shelf and it's very easy to advertise a shape as it is more memorable than just another tall slim bottle, or yet another rectangular box. Examples of pack shapes that have become synonymous with product include the famous Jif lemon and perhaps the most famous shape of all time, the Coca Cola bottle. Even though the bulk of Coke is now sold in cans and PET bottles, Coca Cola always features the bottle in TV advertisements. That sinuous contour is also embodied in the white wisps down the stunning red cans and on PET bottle labels.

International brand identity

Having established a brand together with a marketing hook or advertising theme, the next decision designers have to face is whether or not to extend or globalize that product design. Coca Cola did it through the pack shape and colouring, because the name for Coca Cola differs from language to language. The power of the visual symbol, combined with colour and marketing muscle, is not to be underestimated.

West German designer Studio Kreuser has worked for a number of major international clients developing product brands and packaging, putting well-known names to work in conquering new markets in Germany and Europe.

For instance, Kreuser has been working for Kimberley Clark since 1973 when it supplied designs for the classic German Kleenex box.

The studio's work continued with the development of sub-brands, Sneezies, Boutique and Brevia, clearly labelled with the Kleenex name.

In 1980 Kreuser developed a new West German trademark for Mum roll-on deodorant and redesigned the label around it. The rising sun logo behind the strong, sans serif logoface, helped the deodorant become the market's best seller. The designs were updated twice before the roll-on was complemented by the Mum deodorant stick series, using the same logo. The line is still the market leader.

In some cases advertising is used to educate the consumer in another country about a product, or indeed a pack. The French makers of Orangina designed and ran a series of highly colourful television advertisements about how to shake the bottle and stir the sediment to wake up the flavour of the drink. Another example along similar lines came about when drinks cartons first became acceptable. There was a degree of adverse media publicity about the alleged difficulties of opening cartons in general, so Elopak, the Scandinavian manufacturer of gable-topped drinks cartons, ran a short series of television advertisements showing a man opening a carton. Since then, the cartons have borne a small thumbs-up logo and the words "Elopak Easy-Openers" to distinguish its gable-topped cartons from the rectangular cartons produced by fellow Scandinavian Tetra Pak and the UK-German organization Bowater-PKL, among others.

These few examples show that brand managers and pack designers should not simply look at research in one country, the UK or Germany, for example, but should initiate a continent-wide spread of research programmes to enable sensible advertising and marketing schemes to be adopted.

This Japanese confection, left above, is aimed at health-conscious bodybuilder types, trying to convince them that the bar gives the buyer strength. The strong primary colours reinforce that image and the bright oriented polypropylene pack feels slick, as well as protecting the bar within.

When the product is colourless, tasteless and absolutely without character, like the Perrier shown left, the package effectively becomes synonymous with the product. Note, however, that colour and shape can be imitated by the competition. (Photo Wolff Olins)

In 1980, Studio Kreuser in West Germany developed the new Mum trademark shown here and reworked its label. The product's presentation was subsequently changed again - twice - to take it further up-market, to approximate the look and feel of more exclusive cosmetics.

Corporate identity

For many retailing chains now, the brand identity is part of a wider, corporate identity, on which huge resources can be expended, covering company communications, retail design and packaging design. Packaging offers an opportunity to carry through a corporate design and message.

In the late 1970s and throughout the 1980s, the strength of the multiple grocers and the increasing prominence of DIY and chemist chains caused designers a problem by polarizing the packaging market. In UK food retailing, for instance, Sainsbury, Tesco and four other retailers account for more than half the market, which gives them immense power over what products reach the consumer and how they must reach them. Most retailers, for example, now shun or restrict the amount of display material a manufacturer can supply to aid sales. As if this wasn't difficult enough, the retailers then introduced own-label brands.

To begin with, own-label products were usually commodity products sold at rock bottom prices with a utilitarian design, often all one colour across very different product ranges - from baked beans and fruit juices to deodorants and toilet rolls. Then Tesco and latterly Asda - active also in other West European countries - moved their own-label goods upmarket, with high-quality printing and good

quality photography, vying for more of the consumers' attention and money. The change was dramatic and now brand name products have to fight not only each other, but the retailers own brands, too, on perceived quality as well as on price. In the US, Milton Glaser designed the packaging for Grand Union's supermarket chain, as an integral part of the retail design (see pages 26-27).

Another example, of using corporate identity in packaging design is Shell's repackaging of engine oil. Engine oil has historically been packaged in tinplate jerricans, but Shell decided that it would be a good idea for its own product to be packed in another of the company's products: a plastics jerrican. In this way, Shell believed, its own package could become a brand. This was duly executed and the containers are moulded in the company's own materials, using the corporate colours.

Perhaps one of the most famous retailers in the world is Harrods, which has carefully created the image that it can meet any customer's needs. But a much more important part of the Harrods image is that it is seen as a store from which people like to buy. The greatest product marketed by Harrods is what James Pilditch, author of *The Silent Salesman*, calls 'a sense of affluence, of well being, of social acceptability'. He adds that Harrods has skilfully widened its market, without seeming to trade down. Consumers are satisfied by buying from Harrods, secure in the knowledge that anything purchased from Harrods is acceptable to their peers or to people they try to emulate. The Harrods shopping bag is perhaps one of the most famous examples of identity promoted by secondary packaging (see page 41).

The exciting colours and shape of the French Orangina bottle have been used to the full, even to the extent of being incorporated into a zany television advertisement inviting the consumer to shake the bottle to release the real taste of the product.

Using a postage stamp idea to suggest beverages from far away, the Michael Peters Group designed the concept of a "Fine Foods" range for these own-brand products for the Boots UK chain.

Royal Dutch Shell switched its graded oil products from tinplate into high-density polyethylene containers made from their own products, many pigmented yellow, and bearing the company's shell logo.

Adding "character" by packaging

Because there is so much competition on the shelf in every store, packs have to do more and mean more to the consumer. Every aspect of the pack has to work towards attracting the consumer to buy, and that includes any extra emotive drive the designer can build-in through colour, image and the choice of packaging material.

Designers can, however, do more by adding character or romance. By imbuing a pack with a character of its own, or more accurately a perceived character, the consumer is invited to associate that product with all that is good about any particular romantic theme. A key point about character packaging is that it usually adds value to the product within. This is important because price is no longer considered as the only factor by many consumers when choosing between products. Many discount stores have removed the word "discount" from their names and fascias - the in-word now is quality, and the first and best indicator of that is a product's packaging. Of course, packaging can only secure a trial purchase initially - the quality of the product within is thereafter the key in determining those important repeat purchases. Character packaging can bring an extra dimension to the product, the chances of the consumer coming back for more and buying again are much improved.

By using character in pack designs, designers can take a basic commodity, such as tinned soups and move it up or down market with ease. A product can be made to appeal to a young market or an mature one. The designer can play on traditional loyalties, or make an impact with new ideas and designs. The pack can also exploit the current consumer trends, such as the interest in health or nostalgia for the past.

Health

Health is seen by many consumers as an important consideration in purchasing food and bodycare products. The increasing availablity of fresh foods and delicatessen sections in supermarkets, particularly in the US - which retailing experts describe as five to ten years ahead of the UK in consumer trends - poses a challenge to packaged goods. This has yielded a new character for packaging to pursue in promoting health.

By emphasizing naturalness and the absence of preservatives or additives (see Chapter 4.1 for details of these) with graphic splashes of colour, the designer can tell the consumer that a product is "healthy." Whether customers will believe that packs sporting garish colours and shapes, such as those used across the corners of magazines or in "money off" offers, are in fact healthy is another matter. A better method of promoting that healthy feel might be to incorporate wholesome images and relaxing colours on the pack. It may even be better to allow the product to speak for itself, if it is a food: choose glass or a rigid plastic such as PVC, PET or rigid polystyrene.

Using colour and style

Not all character design need be associated with any particular theme but can also be used, too, to suggest qualities such as crisp and tasty, for snack foods, or warmth and softness, with appropriate combinations of colour and style. Qualities such as softness and temperature can also be conveyed through the actual pack material itself. Glass and tinplate cans enhance the cooling characteristics of the drinks they contain because they actually feel cold. Cartons, on the other hand, do not convey this impression. Plastics can give the impression of being synthetic and aseptic, so are useful and practical materials for pharmaceuticals or first-aid dressings. This last characteristic can, however, be reversed where desired by effective labelling and positioning.

On-pack promotions

One source of anxiety for the pack designer wishing to disguise imperfections in a product by upgrading its packaging should be aware that most modern consumers are no longer deceived by false on-pack promotion. Attempting to promote a mediocre product line in magnificent packaging will always result in a vote of no confidence from the consumer, one that will perhaps affect sales of other company lines.

Nostalgia

Perhaps the most interesting of recent trends is the one taking packaging backwards. Nostalgia has always been important in everything the human race does, but as we move inexorably forwards, headlong into the twenty-first century, with computers that think and cars that talk, we all need sometimes to seek respite in the values of the past. But very often that nostalgic feel is an illusion created by designers to take advantage of this trend, using traditional typography and images from the past, rather than photography or the more modern typefaces.

The designer's work extends even to the shelves and gondola ends as here with the Lifecycle product range, designed by Brand New.

The nostalgia boom has been a godsend to the makers of tinplate containers who have rapidly developed a repertoire of gifts and products for every occasion as seen below. The use of this traditional packaging material from the pre-plastics era also adds value to otherwise standard products, relying on nostalgia for a lost age of quality. Nostalgia has also been combined with "healthy eating" in the use of images conjuring up "farmhouse" or home cooking and produce. An example of society's perpetual myth of a past utopia.

This new perfume, Victorian Posy, was created especially for the Victoria and Albert Museum's Garden Exhibition. The perfume was given a dark green box, decorated in true Victorian style. (Designed by Michael Peters Group for Penhaligon.)

The range of colour-coded labels from UK retailer Tesco, shown left, designed by David Davies Associates, not only informs the consumer but reinforces the company's identity and image as a health conscious retailer.

CASE STUDY BODYSHOP

For the Body Shop, a young but successful UK health and beauty product franchise, with stores in Australia, Singapore and the US among others, packaging is an integral part of its marketing strategy. The credibility and personality of the shop, indeed its very identity and beliefs, are embodied in the understated plastics bottles and jars and consistent labelling approach.

The Body Shop opened with a mere 15 products, each hand labelled by Anita Roddick, the store's charismatic owner/director, who says: 'The cosmetics industry is primarily a packaging industry. By its nature it produces a lot of garbage.' In contrast, the Body Shop's packaging is minimal. No outer packaging is used, no disposable cartons, and products are differentiated by labelling alone. To display useful information for the growing band of environmentally conscious consumers, the Body Shop displays information cards listing ingredients and their functions alongside the products.

Another aspect of the Body Shop's retailing methods, which hark back to the days of small corner stores, is that most products are refillable in shops at a discount. This is partly because of the cost of new packs, but also because the 'recyclability' of the packaging is a selling point. Plastics is also used because of its excellent barrier properties, as well as its durability and robustness. The weight of glass, Body Shop believes, is too wasteful in terms of energy used for transport, especially with small capacity products. The company is, however, working on biodegradable plastics polymers with a small offshoot of ICI. 'When the company comes up with a solution it will be a major breakthrough and we would be a huge market for it,' said Nicola Lyons, head of environmental projects. The only way, therefore, that Body Shop can reduce the effect of extra packaging on the environment at present is to concentrate its efforts on recycling.

By 1991 Body Shop intends that a high proportion of its bottles will be designed and produced by the company's own blow-moulding plant. Here any bottles recovered from customers will also be returned. It is also reducing its reliance on HDPE plastics carrier bags by introducing paper bags at point of sale. These are manufactured from recycled paper and - an important marketing point - carry informative environmental messages. Some plastics bags are still used by Body Shop, but these are made from bio-degradable plastics - another effective marketing point.

Trying to tap into the cosmetics market for men will be as difficult for Body Shop as it is for the other main-stream cosmetics vendors, although this striking design based round the letter 'M' could carry it off.

The company's gift packs have been successful, especially for the festive Christmas season. Acetate packs and ribbons cheer up the plain plastics bottles and wicker baskets from the Far East, packed with wood wool, store a small selection effectively and attractively.

Having extended the range of cleansers and moisturisers so far, the company now has to move into the realms of more traditional cosmetics with this range of perfumes and oils.

From its relatively humble beginnings, the Body Shop now has almost 400 retail outlets worldwide, all managed centrally from the UK. The company's product range appears in a range of five standard high-density polyethylene bottles, ranging from 60-600ml in capacity. The labels are all green, bearing the Body Shop laurel logo and black sans serif type. The whole image coheres well and effectively to persuade the customer that all products are reliable and trustworthy. The Body Shop formula has been so successful that cosmetics suppliers themselves have been quick to use it.

The Body Shop is now trying to expand its product range. Adding to the simple moisturizers and cleansers are special perfumes, aloe vera UV filters for tanning products and - in an attempt to attract the male customer - a range of shaving creams and skin preparations. Unlike the standard Body Shop packs in translucent HDPE, the Mostly Men range is packed in white, with a stark capital 'M' on its side being the main design device. The 'M' image is exaggerated by the bold capital 'M's in the Mostly Men logo, too, but the Body Shop logo and laurel still appears, tilted at a slight angle, at the top of the containers.

1.4   **Graphic design**

Graphic design is an integral part of designing a pack to sell the product, as well as to make it clear and informative for the purchasers. Its first, and arguably most important task is to attract the customer's attention. After that, it must clearly inform the customer what the product is and what benefits it offers, for example, by showing what the likely results of using the product could be - a shiny clean car on a bottle of car wax.

Perhaps the most famous consumer product of all time is shown here in its latest incarnation - the can - using the traditional and striking corporate logo, wound into the familiar Coke contour on a bright red background. There are variations in the style of the pack all over the world, but it remains instantly recognizable, to its rivals' chagrin.

In general, aesthetic and marketing reasons for pack illustrations should not be separated. Should the pack show the product on a pack, or should it show the result of using the product? On Black & Decker's boxed range of lawn tools there is enough room for both: the lawnraker can be seen, but so can the perfect lawn.

Graphic design has a key role in establishing product and brand identity. The most famous example of how colour. shape and typography can be used to do this is perhaps the Coca Cola bottle (and now the can) which is immediately recognizable the world over. Graphics can be used to project similarity, as well as uniqueness: manufacturers attempting to cash in on the success of a product can adopt a similar graphics style to show the consumer "here's another product for that market, but its advantages are X."

As discussed in Chapter 1.3, graphics also play an important part in creating character and image for the product, in conjunction with the pack materials.

However, after the pack has achieved all this, it must then fade into the background a little. Packs should also be acceptable in the home, or in the office, in the garden or the garage. At the point of use, products do not usually need to stand out with as much vigour as they do in the retail environment. This is one of the great dilemmas of graphics design - how to balance the need for the product to be an effective sales tool and an acceptable commodity. It is one, all too often, that designers fail to achieve. In summary, then, the basic functions of graphic design in packaging are:

- to identify a product in the market,

- to inform the consumer about the product it contains and the benefits of using that product,

- to increase the sales worthiness of the product,

- to add character and value to the product,

- to appeal to the consumer long after the purchase is made.

Projecting a message

Deciding what message the product should project is an essential part of the marketing process and the most important task for the packaging designer. In order to make this decision, it is important to ask the question: why should the consumer buy this product? These decisions are most usually made after detailed discussions with the product's manufacturer and the marketing department.

Having decided on the message - the unique selling proposition of the product - it is the designer's task to use every aspect of the pack to persuade the customer to buy. The important point for the designer is that "every aspect of the pack" should include the choice of physical packaging, as much as weight, size or price. But the first point of reference is usually the graphic design of the pack or label.

Unfortunately, there is no hard and fast set of rules that designers can take and apply from one product to another, but they must have some inherent ability in deciding what colours work well together and what impressions lines and shape can create within the mind of the casual passer-by in a supermarket.

From research projects, it is clear that the visual elements of pack design - colour, shape and typography - can enhance or distort the consumer's perception of packaged products. It is useful for designers to be aware of this research into customer behaviour.

There is a film that psychologists use to analyze whether patients respond more to colours or to shapes and forms. The film shows a series of abstract shapes and colours in which the shapes move from right to left and the colours from left to right. From the anwers to the question, "Which way is the design moving?", it is possible to tell whether form or colour dominates in someone's minds.

The test has wide implications for pack designers. Young children, for instance, respond overwhelmingly to colour, while adults respond to form and shape. Men apparently respond more to form, though, than women. There are well-documented differences between different cultures and groups of society, too, which can make designing an international pack difficult.

In this pack of Duplo, by Lego, the bright, primary colours depicting the contents, and a photograph of a child playing with the toy are immediately attractive to a small customer. The carton itself, with its integral handle, could be used as a plaything.

This container for the Black & Decker Lawnraker shows the use of photography to demonstrate the performance of the product.

The designers of the Michael Peters Group were given as their brief the task of creating an exciting personality via graphics for the Brocks brand of fireworks, a product that has to make an impact, being purchased for only one short season a year.

With this range of aerosols, from the Yellow Can Co., the Michael Peters Group has selected a colour to represent each product's household function. The straight sans serif text on a plain background also suggests that the products are cheap and functional.

Colour

Some years ago, a US soap-powder manufacturer gave a set of three boxes of soap powder to various families and asked them to test the products for a few weeks. At the end of the trial period, the families reported what they thought of each powder. The boxes had three main designs: one was bright yellow, another was blue and the third was a combination of both colours, but, unknown to the testers, the powder in each of the boxes was the same. The majority of the families reported that there was a distinct difference between the powders: the powder in the yellow box was too strong (some people suggested it ruined their clothes), but the powder in the blue box was not strong enough.

It apparently left clothes looking dull and dirty. Strangely, the powder in the blue-yellow box was fine.

The lesson of the test and many others like it is that designers should look to use the body of pack design knowledge that has built up over the years to provide a sound basis for graphic design decisions.

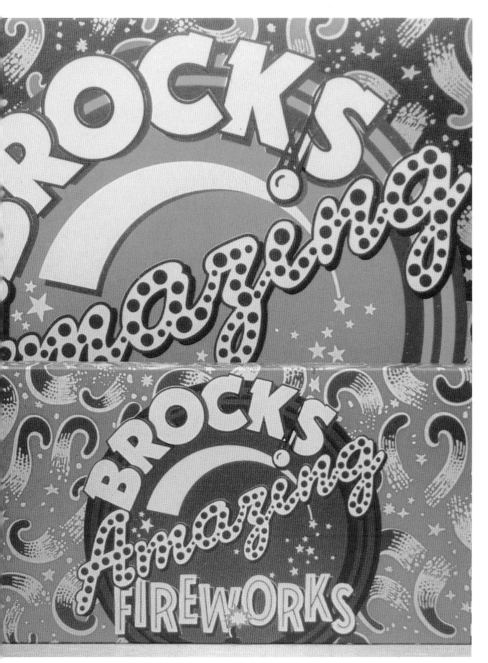

The colours on soap powder packs can have as much effect on dirty laundry as the powder itself! Here, the bright, primary colours and typography of the carton have been carried across to the new plastics container for Persil.

Eye tracking

A more complex method of testing consumers' reactions to designs is known as eye tracking, a test most often carried out in North America, where the retailers in the US's highly competitive market can afford the equipment. Eye tracking relies on a beam of infrared light to pinpoint where a consumer is looking. It records what the consumer looks at first on the pack, or indeed on a shelf. Usually the viewer is shown a series of transparencies or video films of different packs and the infrared scanner is concealed behind the screen. Initial results indicate that bright and light coloured packs score highly, as the eye is attracted to light colours. Loud and obtrusive colours, not normally associated with a particular product category also rate highly. Although the eye-track recorded on computer disc or tape is very useful in determining which designs a viewer "sees" first, it must be consolidated with one-to-one interviews and possibly questionnaires to confirm whether the product would actually have been picked up or bought.

Shape

Shape has always been an important visual element in graphic arts and it is easy to see how it has evolved as an important part of the pack designer's work. It can be used in various ways. First, there is the shape of the label on the pack. There is the shape the items shown on the label take up in relation to one another. And finally there is the shape of the pack itself. For example, square and rectangular packs can be used to good advantage as they offer the greatest face space for the product's message.

In addition, cuboid products stack closely together and waste less shelf space than circular or oval packs. There is some evidence to suggest that retailers prefer these packs, particularly in the fast-moving consumer goods markets, and consideration of shape is extremely important for this reason alone.

If the shapes embodied within the pack do not interrelate comfortably, then the potential purchaser sees a mess of type and colour, and may pass the product by. A great deal of research has been carried out on the subject of shape by the Hamburg design agency, Creative Team, which has developed a test apparently more suited to a psychologist's practice than a design office. The agency says that every pack face falls easily into two vertical halves, with one half dominating the design more than the other. Sometimes a pack appeals to a buyer against his or her will, defying logic and understanding, and the Creative Team believes this can happen because of the relationship of one half to another - the inherent shape of the pack offered to the consumer appeals in some subconscious fashion. The only way to determine whether the shape built in to a pack is acceptable or not is to test both halves, mirrored against the original.

Shape is the key identifier here in this presentation pack of Bell's whisky, designed by the Michael Peters Group. The bottle is shaped like a handbell, with concentric ridges giving the appearance of a handle. The quality of the pack warrants its own gold-blocked corrugated carton outer, in a black satin finish.

In conjunction with another German agency, Schroder/Steinhausen, the group carried out pilot tests in which two different cigarette packs were shown to an equal number of men and women. The test itself was designed to answer a series of questions including how the respondent thought the cigarette would taste and what the quality of the cigarette would be, simply from seeing the pack. The two packs chosen for the first tests were Peter Stuyvesant, which on balance the men in the test preferred, and Kim, which the women preferred. These simple feelings based on pack design alone were borne out by existing product sales.

Size

Another of the visual elements a designer can use is size. Cereal packets, for instance, are usually large, although buyers often complain that they are too large. However, research has shown that the large size of the cereal packet is almost as much a part of the pack as its colouring. It gives the consumer the feeling "of bounty, of expansive energy-giving food" (Pilditch, 1973). Small cereal packs, conversely, would make cereals seem heavy or solid. Another example of the importance of size is occasioned by those tiny gift packs of perfume, which give the recipient a feeling of something precious or expensive.

In considering shape, the designer should also take into account the possibility that the pack may not always be seen in optimum conditions, i.e. on an eye-level shelf under the correct lighting conditions. Around 75 per cent of all Western packaging is designed to be seen at adult eye level, but in an average supermarket, only around 12-15 per cent of the items receive the benefit of good lighting and prime shelf position.

Rectangular bags lying flat - compost in a garden centre or cement in a builders merchant - also project a rectangular shape to the buyer, but from an angle. And the designer must understand the complexities of gusseting and the current state of the art in printing, too, before specifying impractical graphic designs and sizes. The most effective sack designs seem to be the simplest, with few of the intricate touches that can sometimes be seen at eye level on the supermarket shelf. For builders' materials such as sand and cement, strong typefaces - often capitalized, usually sans serif - help reinforce the strong image.

Simple, yet striking colours, combined with illustrations mean that this growbag, designed by David Davies Associates, will attract a buyer's attention on the floor of the garden center.

The "Sfera" theme for this new perfume begins with the spherical stopper and is repeated in the shape and design of the carton, reinforcing the image for the user. (Designed by the Michael Peters Group.)

Typography

First experiences with the basic tool of communication, type, can be confusing. There are so many styles to choose from and so many faces in each group. There are even new faces just waiting to be designed. There is no doubt that the world of type can be complicated, but as with many aspects of design the correct approach opens up endless possibilities. For instance, many designers come through college having worked with only a few faces and then stick to them for the rest of their working careers. More creative work seems to come from people who introduce seemingly incongruous faces that nevertheless suit the job exactly.

However, this particular talent is difficult to learn college-fashion, but comes with experience. So a good tip for newcomers to packaging design is to keep an eagle eye out for text in general - in newspapers and magazines, and on street signs. Very often students are surprised how soon new ideas start to creep almost subconsciously into their work.

The strength of type as a marketing tool can be seen in packs that use different styles of lettering to imbue character or mood in display headings. Gift packs of confectionery, on the whole, adopt flowing text with long thin stems to suggest a mood of sophistication and delicacy. Cartons for a computer's floppy disks tend to use solid strong faces - serif or sans serif - to represent reliability.

Graphics include not only colour and shape, but typography and spatial relationships, too. This range of makeup products was designed by David Davies Associates for the UK Boots chain using different packaging materials but establishing a common identity.

In addition to display headings, the text for ingredients or additional information must also be chosen for readability, and here the standard rules regarding typography will stand the designer in good stead. Text that is centred or ranged flush right is sometimes more difficult to read than material ranged left unjustified, for instance. Line length should also be considered carefully, in conjunction with text size, depth of text and any inter-line spacing or leading. If the lines are too long, the text becomes difficult to read as the eye sweeps on to the next line. Large slabs of italic text are also difficult to read and are so best avoided for technical information on product use, especially on packaging for pharmaceuticals and poisons.

Another question, often forgotten by designers, is "How old are the intended users of the product?" Age, of course, affects eyesight and in general larger type is necessary for products aimed at the elderly or infirm. Pharmaceuticals are perhaps the most important consideration here, but mobility aids and items such as "kneelers" for gardening are not exempt. At the other end of the age spectrum, children need simple and reasonably large type also. Safety warnings - if any - should be prominent, but the typography should be aimed at the adult reader

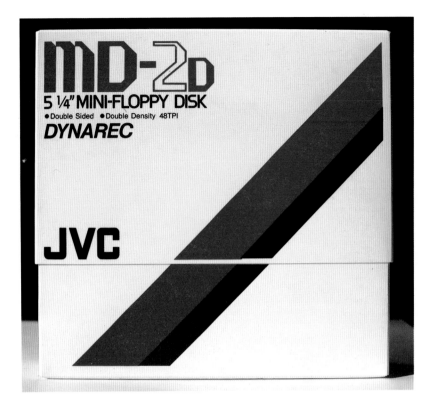

Michael Peters Group's redesign for Winsor & Newton's inks switched to cartons with attractive and amusing four colour illustrations, reflecting the possible end-uses of the product. Ink sales rose by more than 600 per cent.

The choice of colour and typeface by graphic designers can imbue products with a character all of their own. On this computer disk pack, above, for example, the solid sans serif text and bold stripe implies a functional, reliable product, that will store data for many years.

The pack facing

The main opportunity to be exploited by the designer is the pack's facing - that portion of the pack which faces the consumer from the shelf. In many cases this will include a label and, despite the increasing number of legal requirements labels must meet, it is still has great potential for graphic design. In some instances labels can be used as wrappers, physically binding two or more items together.

Where the pack can be printed directly, the label is often dispensed with. In these cases, the pack outer takes on the label's function. Sometimes an outer is not necessary at all. Some designers believe that the use of labels will decrease as printing technologies improve. If full-colour photographs can be reproduced with no loss of quality on any material - from aluminium and polyethylene to zinc, for example - why use a label? However, there are equally clear signs that the convenience and cost-effectiveness of labelling will keep it competitive for some time. Also, the qualities and respon-siveness of paper and board as printing media will ensure that the label will be with us for many years to come. Indeed, the latest technical developments in labelling, covered in the next chapter, are extending the appli-cability of labels still further. Key examples include the self-adhe-sive label, the leaflet label and more recently the in-mould label (see Chapter 1.5).

In pack design a good corporate symbol or brand identity serves well as a means of identification. It also serves to strengthen the bond between components of a range of goods, and enhances the chance that a buyer pleased with one product may try another from the range. The sword is two-edged, though, as it can just as easily put someone off a range entirely if one member of the group does not taste right or does not work as the pack promised.

Photography versus art

The question of whether to use photography or art to represent your chosen illustration also draws arguments in favour of both. One confectionery manufacturer in conversation with James Pilditch, author of *The Silent Salesman*, wanted a design for two ranges of boxes: one was expensive, the other cheap. But the maker could only afford four-colour photography on one - he suggested the cheaper box be two-colour. His assumption was wrong, though, because research then showed that more sales should result from the expensive design on the low-value goods, while the less-expensive, under-stated design on the more expensive box also did well.

Shown left, above, Brand New has relied on an exciting blend of colours, pictures and type styles to give the feeling of the Caribbean. In this case and with the own-brand pasta from Fine Fare below it, the colours indicate the location of the dish. The red, green and white of the Italian flag are incorporated in the border to the central part of the label and around the edge of the pack itself. (Designed by David Davies Associates).

The photograph used on the lower range of cans, redesigned by US Gerstman and Meyers to show a greater variety of use for the product, are much more attractive than the 4-colour illustrations used in the older design and have proved much more successful in the marketplace.

All these developments show - quite strikingly - that the type of illustrations used, and whether the product is shown or not on the pack, all depend on marketing objectives. In other words, the design embodies every facet of the marketing message. How to interpret that message and create stunning pack designs is simply a matter of experience.

Perhaps the time for clear, bold and blatant packaging is past. Perhaps now new kinds of design are emerging, designs that try to evoke a mood about the product in the consumer's mind.

In the swiftly evolving scheme of marketing, more emphasis will fall on the package. "Already the pack is asked to stimulate new product success and put new life into existing products," says James Pilditch. "The package... like any creative salesman... must be used to 'spark the dream, but sell the reality'."

CASE STUDY TOYS

Nowhere is the use of stylish graphics and colour more important and yet more complex than in the packaging of childrens' toys. First, the market for toys has perhaps the most discerning public of all - children have little respect for price. But also there are many manufacturers in the toy-making game. Colour and style are used to force the buyer to give his or her attention.

More than this, the packs must also appeal to parents and other prospective present buyers, so the sales message must still be clear, even through bright graphics and modern imagery. This is especially true of expensive toys, usually well out of reach of pre-teenage children, who tend to have the most exotic tastes.

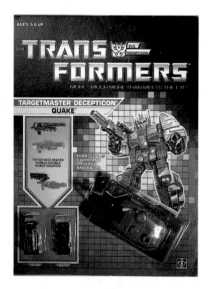

Toys are predominantly playthings, but when designed for bathtime or for the beach or for the garden they are aimed at helping adults enjoy a few moments peace. The transparent polyethylene bag and PVC windows allow the toys to speak for themselves.

Today television or cartoon characters are used to brand products, especially in the United States. International merchandizing deals are being struck daily and cuddly animals on the TV today might be in the stores tomorrow. It seems to be television series that spawns toy spin offs more than films because of the larger audiences they can command over longer periods. The film Ghostbusters, despite being a blockbuster at the box office, only spawned merchandizing when the cartoon series based on the characters, The Real Ghostbusters, hit the airwaves. However, successful television series based around toys have spawned films. The Transformers are an example of this and Masters of the Universe another.

Other categories of toy include the household lookalikes - such as teasets, pots and pans - that get children used to the adult world. Toys that promise bathtime fun are common, too, enticing young children to look forward to the regular grime removal sessions. The packaging for these tends to be functional - often a throwaway bag or pouch with stapled carton header, aimed at ease of access - although the colour and style of these is still aimed at the young consumer.

As with other packaging, the materials that contain the toys must also protect their contents. Corrugated board, litho-laminated for high-quality reprographics are the norm, although standard folded cartons - more often than not employing a PVC window device - are equally common for more delicate toys, usually for slightly older children.

Toy packaging must also be safe. Any plastics must be non-toxic and all corners and edging must be bevelled or rounded. For some days, even weeks, after the purchase, toy packaging functions as makeshift home, bed or garage to both new and old playthings.

Far left: international merchandising deals mean Transformer characters appear around the world. The pack itself is a standard card-backed PVC blister, now fairly typical of toy packaging designs. The logo is the standard Transformers logo, with metallic sheen effect - the colours are still bright.

Left: colour - more than form and shape - attracts young buyers, so designers must focus their attention on being brighter than the competition without becoming garish.

Nostalgia also reigns in the toy packaging world. This marble-effect folding carton, right, embellished with gold banner and fine rules, appeals to the slightly older child with the promise of times past.

In some ways, redesigns face more problems than a straightforward first-time design. This is the conclusion reached by J A Sharwood, which has been supplying Indian foods for almost 100 years. Starting with curry powder, the company's range has extended to include pastes and sauces, a range of ingredients such as spices and ghee, and various accompaniments to meals, including poppadums, rices, and vegetable curries.

As new products were introduced, new forms and styles of packaging were introduced too. To rationalize the natural segmentation of Sharwood's lines, the company decided to redesign the 54 packs in its portfolio. But on looking closely at what it really wanted to do - in terms of what the function of the redesign was - it became apparent that the company was asking a great deal of a simple redesign.

The company's design brief suggested that the Indian range should be seen to enable consumers to create a delicious and authentic Indian meal simply and successfully. The range was intended to appeal to all standards of cook, so the products should be easy to use. The 'unconfident' customer would need clear and concise information on what the products were and how to use them - particularly with regard to how strong curries would be, for example.

In effect, Sharwood believed that its packs should act as recipe books, informing and educating consumers about the possibilities of Indian meals with Sharwood products. What the design should not do, it said, was to overuse already tired Indian devices, such as the Taj Mahal or elephants.

Sharwood also had a problem in that its Indian foods tended to be displayed on the shelf with Chinese and Mexican food as an ethnic mixture, making a greater impact on customers and allowing them time to browse. But in some instances, certain products appeared away from the main family - curry powder was usually sited with other condiments - so the redesign had to work in both situations.

The key to the redesign was a device that would hold the range together. Working with London-based agency, Design Bridge, Sharwood concentrated on the strongest aspect of the product family: the brand name. Having created a working device, attention was turned to the primary use for the products. Each food type was grouped together according to its position. Then the wording for the packs was developed remembering the need to be reassuring, not condescending. The company considered copy for the back of the pack and for the recipe advice, too. This part of the project was important to get right, says Sharwood, before any emotion was built in through graphics.

The graphics adopted by Sharwood included colour-coding for curry strengths and photographs for the convenient hook products such as sauces and pastes, with additional photography of ingredients. The designs were researched using qualitative and quantitative methods to test the impact and imagery and to ensure all the requirements of the brief had been met.

Following the success of the redesign, Sharwood adopted the same procedure to launch its Chinese product range shortly afterwards. The device used here was a Chinese banner in conjunction with the brand name.

Tinned curry: old design at top and above, redesign.

Below, the redesigned Chinese range.

*1.*5

Labels and labelling

Labels and labelling have a long history. Roman apothecaries are believed to have sold herbs in small jars bearing the name of the drug and the seller's name. Wines were sold in marked jars until clear bottles were introduced in the seventeenth century, then labels made from silver or ivory were hung round the bottle necks - a custom still employed today on some whisky decanters, though more for show than for informing the user.

This corrugated carton for Prospero ice cream, designed by Brand New, bears a separate 4-colour label using a hand-written script as opposed to a rigid typeface. The message is "home-made, high quality". In the UK, the ice-cream is almost twice as expensive as any other.

The habit of wrapping produce in paper began originally in the sixteenth century, but no one knows who was the first to mark the wrapper to identify its contents. No one knows, either, who was the first to print a smaller, separate label then stick it on the pack. It was one of those simple developments that was adopted and spread like wildfire across the fledgling retail industries.

The promotional value of the label was not used to any great degree until the last century when French vineyards began to print vineyard scenes on their labels, which had previously been only text. At about the same time, an Irish company, Guinness, began to use the image of a harp to promote sales from their Dublin brewery.

Label design today

However it came about, labels and the labelling process have become increasingly sophisticated, with labels becoming more attractive and cheaper to produce. The trend towards higher quality printing is continuing, enabling increasingly attractive packs to appear on the shelves.

For the designer, more important than the ease with which labels can now be produced, is the fundamental difference in the reasoning behind label design today compared with past designs. The function of the label has changed. Not only must it identify contents, it must also sell them. Consequently, the design and layout of labels, as with other aspects of packaging, is increasingly part of the marketing process. In designing a label, then, the packaging designer needs to consider the same marketing factors as in designing the pack itself, and the appropriate combination of graphics and materials to communicate the product's message clearly.

As with the packaging itself, packaging labels now fulfil a more complex function. They are used to:

- project an appropriate image,

- clarify the identities of product and producer,

- evoke a particular character or mood,

- inform the customer about the product and how to use it.

In addition to the selling function of the pack, the label must often bear some data by law. The quantity of contents in the pack together with the manufacturer's name must clearly be stated, together with a list of active ingredients in some cases, or there should be special warnings if the contents are hazardous. (This is discussed in more detail in Chapter 4.2.)

One of the most interesting of recent bottle labels is this design for Cricketer's Gin designed by the Michael Peters Group. The main illustration is reverse printed onto a self-adhesive transparent vinyl label and applied to the flat rear face, within a simulated acid-etched panel, while only the ball, in red, is enamelled onto the front.

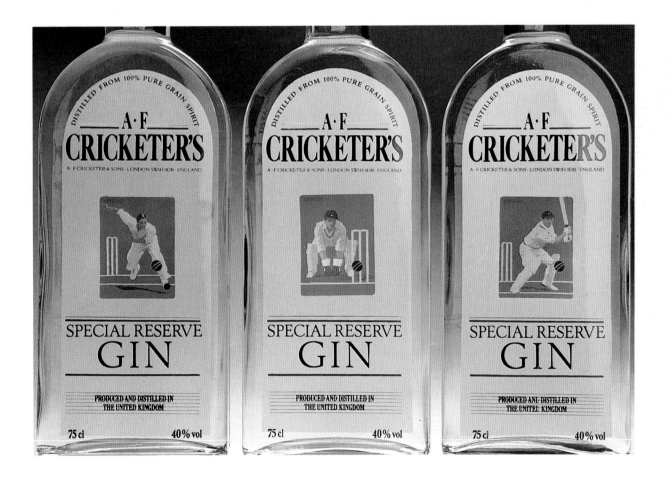

Materials and methods

About 840 million square yards (700 million square meters) of labelling materials are used in the UK alone each year. The latest US figures reveal that 1.5 billion paper labels worth approximately $900 million are used in the United States. The design of those labels is a crucial aspect of packaging design, one that should not be underestimated.

Labels are produced from a wide variety of materials including paper, board, plastics film or sheet, foils, and foil laminates or metallized materials. Metallized foil labels are increasingly used where a quality image is to be promoted, for example, bottled beers, lagers and spirits.

Despite the increasing range of materials used, there are basically only a few major label types and techniques that the designer should be aware of when specifying the label. There are two main techniques for labelling products, using applied labels or direct labelling of the packaging material itself.

Applied labelling includes the following methods:

- plain paper and wet gluing,
- pressure sensitive labelling (otherwise known as self adhesive),
- gummed labelling (usually paper),
- heat-seal labelling (delayed or instantaneous),
- in-mould labelling,
- shrink sleeving.

Direct labelling covers processes such as:

- colour printing,
- embossing,
- enamelling.

The labels on this range of preserves, designed by the Michael Peters Group, adopt a fine flowing script on a black background. The rest of the label is given over to four-colour illustrations of fruit, emphasizing the desirability of the contents.

Colour and imagery play a large part in this label for UK Fine Fare's own brand Rum, also designed by the Michael Peters Group. The nautical element is extremely strong with the illustration of the tall ship and the "sign" flags on a black background.

Applied labels

Until recently, non-adhesive paper labels dominated the labelling market, but now almost half of the world's labels are of the self-adhesive variety. These are obviously easier to apply, although more expensive to produce. John Waddington's Mono-Web label, without the usual silicone backing paper, is a fairly recent development, but it seems to be creating a sizeable market for itself. It took three years to develop and is advantageous from a number of production standpoints. A reel of self-adhesive labels without backing paper contains twice as many labels as a normal reel, enabling it to run twice as long without needing attention. Waddington also believes that the label helps the designer more directly. A wider variety of materials can be used as Mono-Web substrates, and a wider variety of shapes can be cut from the reel. Licensees of Mono-Web have been appointed in the Benelux, Sweden, Switzerland and West Germany.

Virtually all manufacturers of plain-paper labels, including market leader Avery, work within the food and drinks sector, as the following table shows:

The redesign of this wraparound paper label by US designers Image takes away the utilitarian feeling of the original pack. The four-colour illustration with cameo figures leaves Aunt Kitty to the buyer's imagination, while the typefaces are less harsh and the colours less garish.

Application	Worldwide market (%)
● Canned foods	25
● Soft drinks	20
● Beer	10
● Wines and spirits	10
● Bottled foods	8.5
● Other foods	7.5
● Non-food (medical, pharmaceutical, chemical, DIY, toys & games etc)	18
Total	100
Source: industry estimates	

Shrink sleeves

An alternative to sticking paper or plastics labels to bottles or containers is to use a PVC shrink sleeve. In the sleeving process, a machine places a plastics label around a container which then passes through a fairly hot oven-tunnel. This shrinks the label which hugs the container tightly. Some systems can be adjusted so that certain areas receive more heat treatment than others and this prevents the film overheating, which tears holes in the material.

A number of manufacturers offer services in this area, providing PVC shrink labels for a wide variety of container shapes and materials, including glass, metal, wood and carton packs. The labels can also be pre-printed, offering all-round graphics. In these cases, the designer must insist on accurate colour registration of label graphics and barcodes. Shrink sleeves can also be used to form promotional piggyback packs, with trial sizes or free offers of new products. This has been used to great effect in the cosmetics and toiletries market, and to a limited degree in the DIY market.

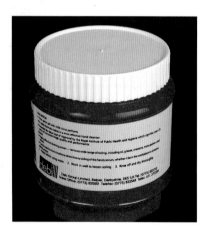

One user of the PVC sleeve is DEB Chemicals, whose Swarfega hand cleaner retails in 250g, 500g and one litre tinplate containers. The main advantage of PVC labels for DEB was the ability to decorate the containers without extensive investment in machinery. The frequent low-volume runs of Swarfega mean that the company can apply the striking red and black labels by hand if necessary.

Direct labelling

In many cases the function of the label has been superseded by the pack itself. Designers can now print textual and graphic information directly onto containers such as cartons, drums, cans and extruded tubes, for example. Where direct printing is prohibitively expensive or impracticable, there are alternative methods, such as embossing and enamelling.

Shrink sleeves have begun to replace paper labels in various markets. The gloss finish certainly adds a lustre to the finished label, but the main advantage over printing directly on tinplate is cost and speed of production. The traditional Swarfega red and white and strong sans serif type are retained in the graphic design.

Enamelling on glass

Although a relatively costly method, vitreous enamelling on glass gives good results and lasts well. The design is resistant to scuffs and knocks. Each colour is applied through a silk screen stencil in a process used in many types of printing. Modern enamels set very hard, very quickly, and a number of colours can be applied in quick succession. The print is set by firing the bottle in a kiln at between 500 and 600 degrees Celsius. This fuses the enamel to the bottle. The range of colours available is wide, and line and tone work can also be handled in this process, opening many new doors for the innovative designer.

Vitreous enamel can be applied as a kind of transfer or decal - a collodion film on a paper backing is applied to the container and the collodion is burnt off during the firing of the enamel. This process is particularly useful for labelling parts of the bottle not accessible to a silk screen, or if more than four colours are required. However, it can be expensive.

Considerable success has been achieved with the development of synthetic colours which do not require firing after the printing process. The colourings are based on epoxy resins and are hardened by various catalysts. These can be applied by silk screening or other methods.

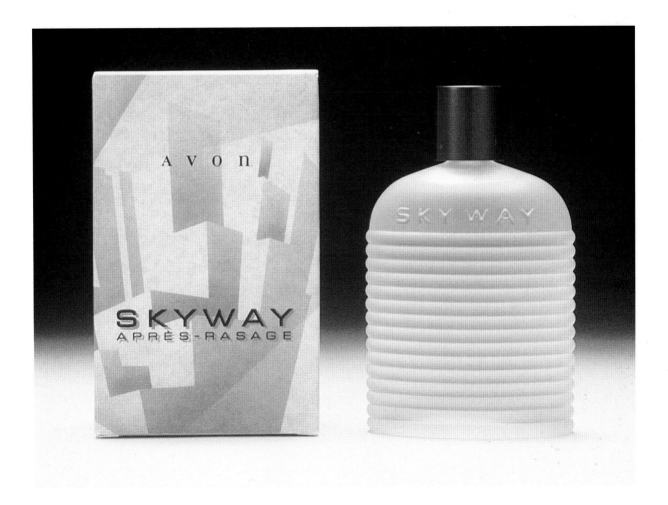

The outer carton for this Avon after-shave cosmetic is printed across its face with the skyline design. Inside, the plastics bottle is moulded with ridges for ease of gripping, and the name of the product, Skyway, is embossed across the top edge of the container. (David Davies Associates)

Embossing

Glass and plastics bottles can also be embossed in the moulding process, which can make labelling unnecessary. Cartons and plastics packs with a clear face can also be printed without the need for a label.

Direct printing

Printing on plastics

The reason for the increasing use and success of direct printing on plastics packs is that the converters of plastics materials have a much more positive approach than formerly to the potential of such techniques. This is particularly true of the flexographic printers, who are striving to match the traditional high quality of gravure printing equipment.

Flexography is not as fast as other printing processes and the quality is not as good as that from a gravure printing cylinder because it uses rubber printing plates. However, it can be cheaper for short printing runs. For long runs - say a million items - gravure is dependable and becomes more cost effective. The colours used by gravure printers remain consistent throughout an entire run, and the process can accurately reproduce very detailed photographic images.

However, the quality and consistency of flexography has reached a point where it is now acceptable to people who want to project a high-quality image at a reasonable price.

Stylish photography combines with subtle typography, with shadow effect, to lift this frozen food out of the ordinary. The marble colour background suggests high-class cuisine. Achieving such a high quality reproduction would not have been possible only a few years ago. (Designed by David Davies Associates.)

Controlling colour

One of the reasons flexography has improved is through electronic colour control. Using a process called under-colour removal, or UCR, electronic scanners eliminate any unwanted primary colours from an image, which helps to prevent colours changing wildly during long print runs.

Another advantage of the modern flexographic press is that it can be used on more materials (substrates) than before. Modern systems can print on cartonboard as well as the increasing numbers of plastics films available today, chiefly polyethylenes and polypropylenes in the foods and industrial oils market. Other films and materials that need to be printed include celluloses coated with a form of PVC called polyvinylidene chloride (PVdC). So, for the designer, the trend means that there will almost certainly be a printing system available for use on any material.

Printing on metals

Similar printing processes can also be used to print directly onto metal packs such as tinplate containers and aerosols. The quality of the image transferred to metal depends on many factors, such as the temperature and humidity in the print shop and the inks and solvents in use. The layer of ink is usually only 2.5 microns thick, so the colour of the ink must be strong and an undercoat is usually necessary. This can increase the cost, but many designers feel that this is justified because of the results that can be achieved.

Offset lithography

Traditional offset lithography has been used for more than a hundred years to decorate sheet metals. (The process is called "offset" lithography because the ink is first transferred from a printing plate to an offset roller or blanket before being passed to the substrate.)

Single colour machines are most usual in this area, so the printed sheets have to be dried before a second colour can be added.

Some two-colour machines have been developed and, despite their expense, these can be useful if the colours on the pack design do not overlap. A close look at aerosol designs will show how designers can use the tinplate or aluminium through the design to achieve the appearance of using more colours (see Chapter 3.4). Another optical illusion is the printing of a second colour over a dried first colour, giving the impression of a third colour. All these points must be considered by the designer, especially as this kind of information is rarely incorporated directly into the design brief.

Wet-on-wet printing

As some designs require more than one colour and need the colours to overlap, attempts have been made to print one wet ink on to another - wet-on-wet printing - so that the first layer is not picked off by rollers during the second print run. However, this method is not fully satisfactory for high-speed, high-class printing, as it cannot be consistently controlled.

Dry offset lithography

A printing technique based on lithography, used extensively in the production of three-piece cans, is called dry-offset or offset letterpress printing. The can is printed by means of a cylindrical plate which wraps around it. Because the plate is effectively a simple stamp, it does not require water. This brings some production advantages, albeit at a cost.

Printing "in the round"

The printing of pre-formed tin containers - usually called printing "in the round" - has been carried out for many years on collapsible and rigid aluminum tubes with up to six colours being applied from one press. High quality packs are usually varnished afterwards in a separate operation.

It is still a good idea to keep the colours separate where possible, but this is difficult and does restrict the design. Improvements have been made to the processes and 150 or more cans can be printed each minute on some machines. With the development of new methods for making drinks cans, high-speed can-printing machines have been developed, chiefly in the US. These machines can handle more than 800 cans a minute, applying up to four colours plus a coat of varnish before the cans move to a stove for drying. Although these tend only to be used by large volume producers with big brand names such as Coca Cola, it is important for the designer to be aware of the potential of the different systems.

This two-piece beverage can printed with strong yet simple eye-catching graphics, designed by Brand New, has done very well since its launch. The effect of the four-coloured box surrounded by text is enhanced on the shelf by the adjacent cans.

The above US range of aerosols, designed by Lister and Butler, uses bright fluorescent colours to identify itself. The use of the word "kills", and the display of dead insects makes it quite plain what the products aim to achieve.

Printing inks

Because of this ever widening range of materials used in packaging, printing inks have also had to change. In the 1970s, PVdC coated polypropylene films were the most popular materials to use and at that time only polyamide inks, or a few unstable nitro-cellulose/acrylic ink mixtures would stick to them. But these all too often melted when the temperature rose a little too high, and they could smell, sometimes affecting the product within a pack. So special inks have been developed to meet the demands of new materials by the ink makers.

How to speed up printing

Various drying methods have been introduced over the years to speed up the printing process: high-temperature air blasts and flame treatments are two examples, but the most successful is exposing the material to ultra-violet (UV) radiation in a process known as curing. The success of the process is easy to explain: it takes about a hundredth of a second to dry a sheet passing through, and even paper, board and plastics substrates now use UV curing. The speed of the process can radically alter production lead times, enabling the designer to meet today's increasingly tight schedules.

A relatively new technique - electron beam curing - uses a stream or curtain of electrons from a heated metal source to dry substrates in a vacuum chamber. The process of generating electrons bears some resemblance to the process of generating X-rays.

Moving away from the traditional solid swathes of colour on paint containers, the line drawings on the new Crown tins suggest that decorating is not so much a chore as an art form.

A major redesign for Crown paints has proven extremely successful in reasserting the brand over ICI's Dulux in the UK. Design agency Wolff Olins also chose to decorate the can lids, enabling identification at lower than eye level.

Heat-sealable inks

One of the more interesting developments has been the appearance of inks that seal pack ends when heated by clamps. Heat-sealable inks first appeared in the mid-1980s, largely on biscuit packs. They make traditional PVdC plastics seals redundant in certain applications, saving time and money in the packaging cycle. However, because these new inks make PVdC coatings unnecessary, the stiffness of the film is lost slightly and the machinability or handling of the film becomes a problem. In these instances a thicker film should be specified by the designer.

UK biscuit maker United Biscuits, owner of the McVities' and Crawfords' brands tested a range of heat sealable inks for its Marie range of biscuits in mid-1986, finding that in general the pack's shelf life was about nine months or so, twice as long as it initially believed necessary.

New trends in labelling

In-mould labelling

Perhaps the most exciting new labelling technique of recent years is in-mould labelling or IML. Developed originally in the USA, IML machines use robot hands to place a paper label inside a container mould just before the pre-formed plastics containers (parisons - like small test tubes) are inserted into and blown to fill the mould. The litho or gravure printed labels usually bear a heatseal coating on one side which is activated by the hot parison. This delayed activation of the label adhesive is important to prevent the labels clogging together in the feeding mechanism.

There are numerous advantages of this method for the designer: there is a high degree of label security as the label is an integral part of the container. It will not usually come off, either during handling, or when the consumer takes the pack home. Also, the label does not wrinkle or crease, and, because it is sunk beneath the surface level of the container, no rivulets of shampoo or engine oil can get beneath the label and lift it away. It also protects the label against moisture and areas of high humidity, such as in the bathroom or kitchen. This keeps the container looking good for a longer period, always a good point to bear in mind in the competition to get the customer to buy the product again.

The lack of wrinkles and creases undoubtedly improves the aesthetics of IML packs and larger labels than would normally be practical can be devised. This, of course, translates into greater marketing impact. On a more technical note, it is possible to reduce the weight of an IML container by putting in less plastics. This is because the stiffness of the label strengthens the sidewall of the container.

IML In Europe

Probably the first company to adopt the technique in the USA was Procter and Gamble, who used IML for its 'Head and Shoulders' shampoo in Europe in 1986. The bottle and closure were designed and manufactured by United Closures and Plastics (at the time part of Owens-Brockway) and the labels were printed and lacquered by the UK's Sanderson and Clayton. On the co-operation of the manufacturers and suppliers on this launch, Procter and Gamble says that, "We always welcome creative ideas from our suppliers. Having evaluated the IML process from UCP, we recognised its significant potential for our business and were delighted to co-operate with them in this venture."

One of the first in-mould labels to hit UK and US shelves was Procter & Gamble's Head & Shoulders shampoo. The label is placed inside the bottle mould by a robotic gripper and the molten plastic material then holds it in place. The advantages of the technique include its speed and that the label cannot be unstuck, ruining the appearance of the container.

Leaflet labels

Another important trend in labelling is epitomized by the leaflet-type of label known as Fix-a-Form from Denny. These are often used where there is a legal requirement to supply more data than can reasonably be included on a standard label, with agricultural or horticultural products, for example, where the product must be prepared or used only in specific ways. Some pharmaceutical products also use these kinds of label.

The future for labels

Labels will continue to be demanded by designers because of their versatility, especially for promotional packs and items on short production runs. Their greatest threat comes from the decline of traditional materials or media, such as glass jars and cans, in favour of new composite, flexible and plastic containers, which can readily be printed using flexographic or gravure methods. But there are new developments, including the pre-printed foam or film sleeves such as United Glass's Plastishield label, now extensively used on both glass and plastics bottles.

2.1   **Paper**

The use of paper as a packaging medium has a long history. Probably the simplest and oldest form of paper pack was a sheet of treated mulberry bark, prepared by the Chinese in the first or second century BC. The Arabs captured the process a few hundred years later and scattered the technique across Spain, France and the rest of Europe. Although paper and, of course, board are not the only or the strongest plant products ever used for packaging, these are the two that have survived the longest. Wooden crates are still used for the export of large heavy machinery, but for anything smaller its weight is against it. Cloth and other textiles were used as sacking for centuries until paper replaced even that. Together with cellulose, paper was the most widespread packaging material in general use - either in sheet or bag form - until well into this century.

Packing fresh produce is one of the most familiar uses of the formerly ubiquitous brown paper bag.

In the second half of the twentieth century, paper has been increasingly superseded by plastics: the ubiquitous grocery shopping bag is no longer brown paper but a plastic bag. However, paper has maintained an important presence because of what is offered by its particular characteristics and is now poised to take advantage of the growing international concern about the effect of non-degradable materials on the environment and the need to conserve non-renewable resources.

These concerns have halted and reversed the decline in the use of recyclable paper for packaging. One reason for the reversal of this trend in Italy occurred in the mid-1980s, when plastics bags were discovered in the stomachs of beached sperm whales. The Italian government promptly decided to ban the use of plastics bags for packaging by the end of the 1990s. The ban includes all packaging materials and containers that are not biodegradable, so its effect will be drastic on designers who need to use plastics for packaging, as well as affecting the output of the Italian's own plastics industry. The ban will also severely restrict exports from the rest of the world into the country. The ban shows how ecological issues have reached the level of national and international policy and, for the moment, there seems to be a swing back towards recyclable materials, such as paper and board.

Competing technologies of paper and plastics

Technological advance and the increasing use of plastics have posed problems for the traditional paper and board-based sectors of the packaging industry, which have certainly declined to some extent. But there is also evidence to show that, following a certain amount of rationalization and consolidation in the late 1970s and early 1980s, when paper-based packaging forms faded from the scene, paper is enjoying a steady popularity. Designers are able to use plastics laminated or coextruded with paper sheet to provide new products with their own special characteristics. The various types of antistatic sheeting for packaging electrical and electronic components are a good example. Some paper suppliers are actually experiencing growth in their business.

To wrap goods in paper today requires a certain effort. The consumer must protect the products which can not effectively protect themselves - and for this reason the gift market makes extensive use of paper bags, printed with crests and using colours not used on other packaging. Ferrero Rocher uses crimped paper cups also to support its expensive foil-wrapped chocolate.

Papermaking

Virgin paper looks something like blotting paper, but through using different processes and by making deliberate additions to the woodpulp mix during paper making, different papers can be produced. Besides relatively simple changes, such as changes in colour by the simple addition of dyes, paper can be made to resist moisture. The most common additive is a natural resin known as "size", which - together with aluminium sulphate - provides that wet strength. Other additives - such as grease-resistant substances - may be applied at the surface, but the commonest basic coating is starch, which seals the surface of the paper and increases its strength. For the pack designer, the strength of paper is perhaps its most important characteristic.

Kraft paper

This coarse paper, noted for its strength ("kraft" is the German word for strength), is an important paper packaging material, used frequently for brown grocery bags and sacks. Most of this is produced in the US, Canada and Scandinavia. Naturally brown, kraft paper can be bleached to a variety of beiges and even to white. A similar paper, sometimes called shipping sack paper, is used to construct multiwall bags and sacks, having from two to seven layers. A plastic or foil layer can be substituted for paper in special cases, for instance, where an additional barrier to moisture is required. This multiply construction makes bags particularly strong and useful for products such as dog food and animal feeds, charcoal briquettes, cement and other building materials. They are also used for fertilizers and agricultural chemicals, but - as always - the product determines the pack. The designer needs to consider carefully the product's characteristics before choosing paper as the packaging medium, for example, it may not be appropriate for very moist or sharp-edged items. Corn syrup solids, for example, need a moisture-resistant bag, while insecticides and other poisons need to restrict the passage of air as they need to be odour resistant. Cement, in comparison, and other dense or sharp-edged products, such as gravels and some forms of charcoal, need strong sacking materials.

Bleached papers
Bleached papers can be used in bags or other paper-based products where appearance and the protection of pack contents are important. Machine-finished (MF) or machine-glazed (MG) papers can be used to make bags or wrappings for bakeries and fast-food outlets, among others. Tissue papers are made in a similar fashion and can be glazed or unglazed, depending on the requirements. A special form of bleached paper is used to make labels: the paper is coated with a fine china clay powder to fill in the normally rough surface cavities. This makes the label surface easier to print and makes the finish glossy and more attractive.

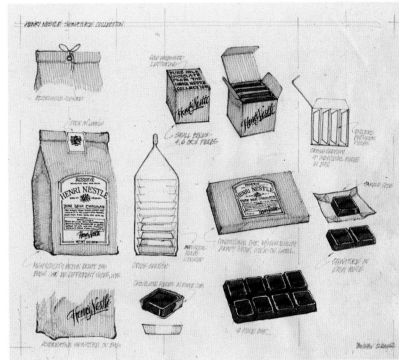

In this, the preparatory work for the Henri Nestlé collection shown on the opposite page, notice the constant use of the phrase "high quality" in the packaging specification.

A strong paper pouch that stands upright because of its flat-bottomed construction is here printed with a crest, metallic gold and a swatch of colour reminiscent of a blue riband. (Designed by Lister and Butler)

Paper has long been used to wrap foods - greaseproof or waxed papers, especially so. With this particular product range, designed by Wolff Olins, Dairy Crest is aiming to give a traditional image by making the butter cylindrical, rather than rectangular.

Paper for food packaging

Although the application of paper in packaging can be used for a wide range of products, as with other packaging materials, food packaging accounts for the vast majority of uses, as is demonstrated by the table opposite. There are several important guidelines for the designer considering paper packaging for food items. For general food use, glassine and greaseproof papers should be used, as these are grease and oil resistant and offer a barrier to unwanted odours and moisture. Waxed papers can also be used for most foods, as they are tasteless, odourless, non-toxic and inert. Fatty foods, in comparison, require high-strength vegetable parchment, which is resistant to the typical staining caused by grease.

End uses of packaging papers worldwide		
● Food		81%
condiments, spices and sugar	*30%*	
bread	*10%*	
confectionery	*10%*	
cereal wrappers	*8.5%*	
fats and dairy produce	*5%*	
snacks and crisps	*5%*	
pet foods	*4%*	
beverages	*3%*	
convenience foods	*2%*	
ice cream	*1.5%*	
dried milk and baby foods	*1%*	
biscuits, cakes, meat, poultry and frozen food	*1%*	
● cigarettes		5%
● medical and pharmaceutical		5%
● paper products (napkins etc)		5%
● detergent, shampoos and toiletries		3%
● chemical and agricultural		0.5%
● other		0.5%
● Total		100%

Source: industry estimates

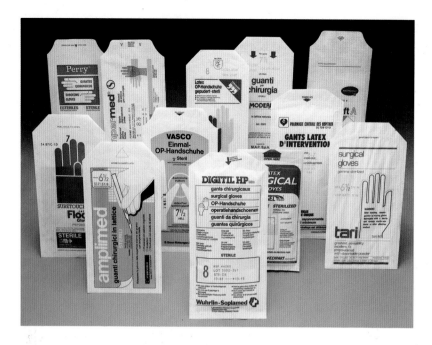

The design of medical packaging

The design requirements of medical packaging are relatively straightforward: the contents need to be protected from the environment, and easily accessible and identifiable. Flexible materials are generally cheaper than rigid containers and they are usually easier to open - either by peeling a self-adhesive strip, or by tearing across the lip. They can easily be sterilized by gas, steam or radiation treatment and are also easy to dispose of. Transparent plastic windows for rapid identification are another useful design feature.

Obviously the criteria for each design factor differs, depending on the client, but in general these few guidelines serve the needs both of large customers, such as Smith and Nephew, or Johnson and Johnson, as well as the the small, in-house sterilization departments that most European hospitals use.

One important design trend has been the adoption of "long-fiber paper", which has no lacquered coating, but can still be sealed simply by clamping between heated jaws. The paper also peels cleanly, making retrieval of the contents particularly easy for a surgeon or nurse.

The UK is a major supplier of medical packaging expertise and companies such as DRG have strengthened their standing by opening up subsidiaries in Belgium and the US, enabling designers outside the UK to specify some of the exciting new materials increasingly available in this field. For the US or continental European designer, paper packaging specialists, such as Gothic House Incorporated, Bonar, Pakcel and Akerlund and Rausing are worth investigating, as well as Senegewald, Finnconverta and Safta.

Medical packaging

A smaller but significant application of paper in packaging is its use in hospitals or in first-aid kits, for which it is an essential packaging medium, being used in the design of 80 per cent of hospital utensil packs. As such, it is unlikely to be replaced as a medical pack medium for many years.

Paper bags used for medical packaging have different porosities and plies and use special folds at the lip to prevent bacteria entering. The bags need to be porous to allow steam, gas or radioactive particles to enter and sterilise the contents. Papers with small holes - no larger than 0.3μ - work well. The smallest of the troublesome pathogens are 1.0μ across.

Many of the papers used are, coated with polyvinyl acetate (PVA) or some other lacquer to strengthen the material against bacterial attack. This does, however, increase the cost of packaging significantly. Where plastics films are used - perhaps where the designer needs to ensure the contents can be seen - paper almost always appears on the other side of the pack. Packs made completely of plastics or cellulose are rare.

Medical paper packaging can be used in cut-sheet form, as well as in pre-made bags or pouches. Reels of paper tubing are also common.

DRG Flexible Packaging, an international market leader in paper packaging, supplies a range of sterilizable packs for surgeon's gloves. Paper is the base material for such packs, but its specification and treatment is carefully varied according to the bacterial barrier performance required in each case.

Paper bags and sachets

The words bag and sack are usually interchangeable, although some manufacturers distinguish between them by the use of an arbitrary weight limit by which "bags" contain no more than 25lb (11.5 kg), while sacks contain anything over that weight, up to the huge wrapping sheets for trimmed lumber used by some sawmills.

The paper bag is perhaps the oldest and one of the cheapest forms of packaging available. It is secure and dust-tight when sealed on all four sides, and automatically takes up the shape of the product it contains. Products such as nails, or grain, which settle on standing, take up less storage space in paper bags because the bag settles, too. However, this characteristic is also a disadvantage: bags do not stand neatly on the retailer's shelf without some form of reinforcement or support.

Most suppliers of sachets for the soup and sauces market, for example, now produce outer cartons which double as transit packs. These hold the sachets upright in the store when the top and front flap are ripped back.

In addition, the wrinkles and folds that some paper packs show can be unattractive and for this reason products that need to be seen as up-market are not normally packed in bags. This homespun appearance can be used to advantage, though, with nostalgia products such as confectionery, spices, garden chemicals and plants.

The flat or square-bottomed bag is quite common in the US and snack foods such as popcorn and crackers appear regularly in this form of container. With this new snack cracker for the UK, Scooples, the maker was planning to capitalize on the success of other foods in the same pack. The use of the pack is a way of familiarizing the consumer with the product before it has even been seen. (Designed by Brand New)

Styles of bags

For thin products, the flat bag or envelope is most economical, but in general one of the other forms is usually required. Gusseted bags, like the traditional grocery bag, which opens out, are useful for bulky contents.

There are four major styles of bag:

- automatic bottom, or self-opening style (SOS)
- square or pinched bottom
- flat bag
- satchel bottom

Paper sacks

Multiwall paper sacks are usually of two forms: either sewn across the top and bottom, or pasted. In both cases the side seam is glued. The sewn sack is usually the cheapest, but it is not as secure against moisture as the pasted sack. When presented to the packer, sacks are usually open at one end or "valved", with a small corner opening, through which the product can be poured. In the past few years, sack-filling machinery has improved and sacks are filled faster than ever: 12 to 15 or more can be filled by one machine every minute. However, as the speed increases, the sack construction itself can impose limitations on the speed of fill. The major problem is letting air out - known as "de-aeration" - so most sack makers now offer a variety of seams and vents to allow designers to tailor sacks to their customers' requirements.

The designer working in the sack market soon gets to know which forms work best from both the functional and the presentation viewpoints.

In combination with other materials, such as plastics and foils, paper sacks can contain up to around 110lb (242 kg) of product securely and with little wasted space. They tend to be used for free-flowing granular or powdered products such as plastics resins, agricultural seed and animal feeds. On their own, they allow potatoes and other fresh root vegetables to breathe, and they are less prone to puncture or splitting than people imagine.

Cement manufacturing has also been a traditional use of paper sacks across the world, largely because the manufacturers have built their own packing stations on site. In addition, cement is packed while it is still hot and few plastics sacks can be used for hot-filling. Milk powders for animal feeds are also packed while hot, so its designers also favour paper.

Sack finishes

For the designer who chooses paper sacks as a medium, as with other paper for packaging, an increasing variety of surface finishes are available to enable two and four-colour printing to be carried out. Many of the mail-order companies are providing printed bags and sacks as an attractive means of self-promotion. The multi-million dollar mail order business in the USA seems set to stay with paper, despite in-roads by plastics, largely because of the traditional look and feel of paper sacks.

Although facing severe competition from the new plastics, sacks laminated with other materials, such as aluminium or polyethylene, are proving tremendously popular, as a replacement for the rigid tea chest. Paper still has the advantages of its stiffness, which enables it to be formed into sacks extremely quickly, and its permeability to gases. This last factor makes it still the most suitable material for packing sugars, although an increasing amount is now going into plastics film.

Interestingly, an important factor that critically affects sales of the paper sack, is the weather. If there is a bad potato crop one year, or if grain yields fall, then no one needs as many sacks. Conversely, if the weather is good and the grass is lush, then farmers do not need quite so much feed, so fewer sacks are needed then either. Save for the soft drinks industry, no other retail sector is as affected by the weather.

The outer carton for the tea, left, for the German market uses black and gold to heighten the quality of the product. In addition, each tea bag is individually wrapped, adding still more to the idea of quality. (Brand New)

The bold blue logo and name of the Blue Circle Cement Company are clearly visible on all faces of these paper sacks, however they are stacked.

End uses of paper sacks worldwide	
● cement and other rock	24%
● potatoes (not small packs)	20%
● feed stuffs	17%
● food	15%
● chemicals	8%
● refuse	5%
● other	11%
● Total	100%
Source: industry estimates	

Cellulose

Strictly speaking, this chapter should also cover regenerated cellulose, or Cellophane, which is also derived from wood pulp. However, as a transparent packaging film it competes directly with plastics films and sheet, rather than paper and paperboards, so it is dealt with in the next chapter.

*2.***2** **Plastics films**

The development of synthetic polymers arose from studies of the known natural polymers at the turn of this century. One of the first synthetic polymers of any direct use in packaging - and still used by today's designers - was the shiny, crisp cellulose acetate. One of the first commercially available moulding plastics was celluloid, formed from cellulose nitrate and camphor, which is still used to make table tennis balls, among other things.

Although cellulose products are in truth derived from wood pulp, rather than petrochemicals, the resultant material - transparent and strong - competes mainly with plastics, which explains its inclusion here. Plastics films and how designers can use them form the major part of this chapter.

Synthetic and plastic films are made from long chains, or polymers, formed from repeating groups of the same molecules, known as monomers. There are various naturally-occurring polymers, such as the protein keratin that makes up human hair and skin. Rubber, too, is a polymer, and so are silk, wood and cellulose. The type of building block used and the way the blocks or links are put together can have quite an effect on the appearance and properties of the material. The designer needs to be aware of the characteristics of the different films and how they can be exploited.

The archetypal plastics bag is actually made from cellulose, a plant product. This shiny film has an excellent gloss and twist-wrap properties second to none. Today, oriented polypropylene can approximately match it, quality for quality. This colourful design is by US agency Peterson and Blyth.

Cellophanes

One of the first widespread and practical packaging films was made from regenerated cellulose, better known as Cellophane, now manufactured in more than 25 countries round the world. Initially, this was only used as a wrapping material, but a coating of cellulose nitrate made the film less permeable to moisture. Perhaps more importantly, the coating meant that bags or pouches could be made from the Cellophane because two sheets of the film could be welded together by heat. Cellophanes are used widely in some of the special laminates and coextrusions discussed later.

The high shiny gloss of the Cellophane surface makes the film an excellent wrapping film for quality goods, particularly where its transparency can be used to good advantage. In addition, the film's deadfold characteristics - the way it stays folded when creased - make it ideal for use in confectionery as a "twist wrap". Boiled sweets, sticks of rock, mints and other sweets still use a great deal of Cellophane, despite competition from other materials, such as oriented polypropylene (oPP). Designers can also emboss the film with patterns, adding yet more value to the product.

Cigarette packs are one of the major non-food sectors still to use Cellophane wrappings, usually now with some form of tear-tape to enable the pack to be opened more easily. Many designers, however, feel that Cellophane packs do not need tear-tapes as it tears readily anyway. It is this tearability that makes Cellophane useful for single-serve or portion packs of sugar and condiments.

Besides the more familiar transparent film, opaque and coloured Cellophanes are now available. The films can easily be printed, using most techniques, although this is often restricted to simple graphics advertising cut-price offers or sales incentives, allowing the pack beneath the film to carry the bulk of the sales and marketing message. Why render a sparkling film dull by covering large areas with print?

Cellophanes, sometimes with micropores to allow bread to breathe, have almost entirely taken over from greaseproof papers in the bakery department, as here in the UK Marks and Spencer store.

Versatility of plastics

Most of the new products and developments in packaging technology and design are in the field of plastics, largely because the few remaining limitations of working with plastics have now been overcome. In the past, for example, carbonated beverages proved difficult to pack in plastics because of the pressure imposed on the container; so too were products that are processed at high temperatures. Now there are trays that can be microwaved or used in a conventional oven. In addition, there are packs that, like a kind of skin, allow their contents to breathe the air.

As an indication of the potential of this material for the designer, it is expected that, in the early 1990s, plastics will outrank glass as a packaging medium in the USA. They will probably overtake metals by the turn of the century. In Europe, plastics are expected to account for more than 50 per cent of packaging by the year 2000. At present roughly a quarter of all packaging is designed in plastics. Although there are serious environmental considerations, research is being carried out across the world to make the plastics package as effective, easy-to-use and economic as the numerous materials it replaces.

Plastics films

The phrase "plastics packaging" covers a group of materials composed of the synthetic polymers described above. These materials can be blown up like balloons or moulded into different shapes to function as containers, or drawn into films of varying thickness, depending on the required function. The basic plastics used to make films are:

- low-density polyethylene (LDPE) and linear low-density polyethylene (LLDPE)

- high density polyethylene (HDPE)

- polypropylene (PP)

- polyvinyl chloride (PVC)

- polyethylene terephthalate (PET) - used in polyester films, for instance to make videocassette tape, but mainly used for containers

Sachets made from plastics films such as cellulose acetate offer a solution to packaging delicate spices and herbs, such as this saffron from Spain, and are transparent so that the the buyer can see the product.

Oriented polypropylene (oPP)

The major competitor of cellulose films is oriented polypropylene (oPP), a white and opaque film produced in large quantities by the Benelux countries. Kalle, British Cellophane Ltd (BCL) and Mobil are also major producers. Designers have found oPP particularly useful for the biscuit, snack and confectionery markets because of its air-tight nature when sealed, together with its sparkling appearance and printability. Major confectionery products including Mars and Aero have moved away from paper wrappers to oPP films. Cadbury's confectionery packaging buyers say this is largely because the film does not show wet or greasy marks on the outside.

An aerated chocolate bar, Aero, taken over with Rowntree Mackintosh by Swiss giant Nestlé, is extremely sensitive to moisture in the atmosphere and must also be protected from the odours of strong-smelling confectionery items and tobaccos, which might penetrate the wrapper, tainting the flavour of the somewhat delicate product. Nestlé say that this has been achieved, using oPP.

The material handles well on packaging machines and provides a good base for printing. Other major confectionery lines to have moved to oPP include Picnic and Double Decker in the UK and Chocolate Caviar from Dearborn in the US. The film has also proven popular with Spanish and Australian designers as their climate demands protection from damage in transit and from moisture.

UK Allied Bakeries has chosen a high-clarity oPP from from MCG Venus Packaging for its Vitbe Raisin Brans. The photographic illustration of the product's natural ingredients is printed in six colours by the "Superflexo R" process specially developed by MCG Venus.

Stronger, thinner films

As with other films, a major trend worldwide is the reduction in film density. Stronger, thinner films are the result; cheaper materials and processing are the goal. These developments affect designers everywhere, as graphics on some packs could become more difficult to print, or the traditional crimped end could become too floppy to be effective as part of a high-quality pack design. Designers should experiment to develop the best effects, but always confer with materials experts before going too far.

Cellulose acetates can make attractive overwraps for luxury items such as here - for spectacle cases from Giorgio Armani, designed by the Michael Peters Group.

Snacks and crisps

Transparent oPP is also used for snacks and crisps, although new metallized oPP films are becoming very popular in this area (see Chapter 2.3). This film can be used by the designer in exactly the same way as ordinary plastics - it handles and seals well - but because the film provides a very good base for graphics, designers should push it to the limit by incorporating four - or even six - colour work where possible. The snacks and confectionery markets are increasingly competitive, worldwide, so the most attractive designs will invariably be the more succesful in the marketplace. Technically, too, these films are becoming increasingly acceptable: manufacturers claim that they increase the shelf life of crisps from about 10 weeks to 28 weeks or more.

Lack of deadfold

The appeal of oPP is not global, however and its use should be considered in the light of market research, such as that carried out by Cadbury's packaging department. This showed, in the case of the Cadbury Flake, that although the product might benefit in cost and longevity terms by changing to oPP from Cellophane, the product's consumers like the yellow twist-wrapped pack so much that sales would probably decline if the company made the switch. OPP's lack of deadfold makes the material difficult to fold and crease, and this could be described as a significant disadvantage. For this reason Cadbury believes that, for the foreseeable future, moulded chocolate blocks will probably remain in foil, with paper sleeves or band wraps. The company is, however, investigating some metallized films which have deadfold properties.

Transparent polypropylene is another form, used in the packaging of stockings and tights, as well as other fashion items. However, without adhesive strips, the material is difficult to seal and tends to be twice as expensive as similar packs made using polyethylene. For these reasons polypropylenes only tend to be used with premium products that can absorb the production costs.

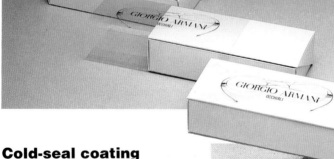

Cold-seal coating

Plastics films are traditionally sealed by crimping the ends and melting or welding the sheets of film together. Obviously, packs for chocolates and candies cannot be sealed in this fashion because the high temperatures would also melt the chocolate. Coating confectionery films with polyvinylidene chloride (PVdC) is the usual solution and enables the packs to be sealed by the pressure of the crimping jaws only. This PVdC coating is therefore known as a cold-seal coating and the process can be expensive.

Polyethylene

The workhorse of the plastics design world is polyethylene, which accounts for around a third of all plastics packaging world-wide. Of this more than 35 per cent occurs in the form of plastics film and sheet, its most common use being for bags and carrier bags. Different densities of film can be chosen by designers to suit various applications. The density of the film affects certain key properties such as film stiffness, resistance to low temperature and strength.

There are two main types of polyethylene - low or high-density - but manufacturers can also mix and match the components of a film, to offer specific properties to a buyer. Linear low density and ultra low density films also exist. ULDPE is said to be 200-300 per cent stronger than low density polyethylene.

High-speed films

The designer's situation is made slightly more complex as the exact choice of plastic requires a precise knowledge of how the films function under different conditions. For example, over-wrapping paper products requires a stiff film for use on high-speed packaging lines, but designers must still offer a clear film with a bright surface shine. A medium density polyethylene is perhaps the best solution. Ice bags and frozen food packaging, in comparison, must perform well at low temperatures. Low-density films could be used, but do not have the right characteristics, so a slightly different material should be used.

Plastics technologists have all the information at their fingertips, often on computer databases and it is to these specialists that the pack designer should turn for more details of high-performance plastics and how to incorporate them into effective and modern designs.

EVA Copolymers

In the polyethylene family, ethylene molecules are linked or polymerized with many other ethylene molecules, forming long chains. But ethylene can also be polymerized with molecules of a similar compound, vinyl acetate, to form ethylene vinyl acetate (EVA) copolymers. These have better properties under low temperatures. For foods that require a high barrier - against grease, water or gases for instance - high-density polyethylenes offer the optimum solution.

A material that the designer could consider instead of EVA is ultra low density polyethylene, devised by the Dow Chemical Co and marketed under the tradename Attane. However, the resin is about 20 per cent more expensive on average. Its main advantages are that it is said to be two or three times stronger than LDPE and it has arguably better optical properties - shine and clarity, for instance. These are a gift for designers in the plastics film market.

Dow claims that because of ULDPE's strength, thinner films can be produced, thus negating the added cost of the raw resin. Meats and cheeses are the first products to be packed in the film.

Resolving the technical problems

There is no doubt among the world's top designers that the flexible plastics film will eventually replace many traditional packaging materials. Polyethylene, for instance, keeps water vapour in - or out - making it useful for fresh foods. It is also suitable for use as a tamper-evident shrink or stretch film. Polypropylene, too, is light and relatively cheap - now manufacturer's design teams are working on using it as a fresh produce packaging material as well.

PVC is being used as a film to wrap the increasing number of fresh food trays in the modified-atmosphere packaging process and polyester films are used for bag-in-box packs and for microwave lidding films. Increasingly these are being decorated too with exciting graphics as well as processing or cooking information.

In addition there are a number of transparent polyamide films, not covered here, that are an effective barrier to gases as well as being easy to print. Their main uses are for cheese, meat, coffee and boil-in-the-bag foods. The US Food and Drug Administration recently released this material for use with dry food and acidic products, so designers should be looking at this material in addition to the standard workhorses.

Another recent development, polycarbonate film, is said to have good physical properties and withstands heat. Probably its first use in the UK was as a pack for part-baked rolls and French sticks for frozen food retailer Bejam.

The ever-increasing quality of flexographic printing for commodity items such as these frozen foods, left, means that more detailed artwork can be used to decorate the products. (Designed by the Michael Peters Group)

Cellulose, which tears readily, is nevertheless strong and can be used to pack oddly-shaped, sometimes sharp goods in a variety of shapes. This range of pasta forms from Fine Fare employs the material to good effect and takes the Italian colour scheme devised by David Davies Associates well.

The designer of tomorrow will be working with all these materials as a matter of course and will need to understand how they can be exploited to produce a successful pack design. Increasingly, the graphic design of a pack will have to take account of the properties of the highly complex materials and the production technologies. The packaging designer needs to consider all stages of the packaging process, ensuring that the material chosen is compatible with the product, especially with foods, and that the packing process provides the conditions required by the product and the packaging itself. Medical goods, for example, produced in aseptic conditions must be kept sterile during the packing process. The compatibility of packaging with foods, the available packing equipment and the degree of protection required all play their part in making a quality pack, even if this is never perceived directly by the customer.

Protection, for example, will vary depending on the fragility of the product and its required shelf life. Perishable food such as bread and meat can be attractively packaged and still have quite a long shelf life with developments in film technology and in modified or controlled atmosphere packaging. The benefit of these processes is often not seen directly by consumers, but retailers can offer better quality products by using them.

Bags and pouches

Plastics bags are very different from paper bags, particularly in the way they are manufactured and in the products they contain. Unlike paper bags, plastic bags for example may be made as a continuous tube, which is sealed and cut at intervals. In some cases, a web or sheet of film from a reel may be folded and heat sealed at the sides to give what is known as a side-seam bag. The folded edge forms the bottom, which collapses like an accordion to give a gusset. The top edge usually has a lip to make it easy to open. This is one important design advantage of the side-seam bag over other bags, which must be cut flush at the mouth.

Bags can be made and stored flat for transport, but the most economical way to make pouches - sealed on all four sides - is to form them at the point where they are filled. Polyethylene pouches can be used for fresh milk, as in Canada for instance, and this type of pack, which can be decorated, has a number of advantages over bottles of glass or plastics. First, it takes up less room. It is also economical, hygienic and takes up less space in the dustbin.

However, outside parts of Canada, where for some time there was no alternative milk container, this pouch has had little success. This seems to be because consumers believe it to be inconvenient. Leakage has caused some problems with these containers, but the manufacturers believe this to be controllable through designing the packs in stronger plastics and using more effective, gusseted designs. And though the potential for the pack is high, with some interest being expressed by fruit juice packers and motor oil suppliers among others, the marketing task faced by designers and suppliers is enormous and the pouch is unlikely to be a major consumer success. This may be for various reasons, including the traditional association of milk with rigid bottles or cartons, but also there are ecological concerns about recycling and disposal. This is an interesting instance for the designer of customer preference being more relevant to the choice of materials decision than purely practical design considerations.

Plastics materials can now be devised that suspend the maturation process in food. These processes such as sous vide (shown below), vacuum and modified atmosphere packaging all extend the shelf-life of fresh goods. (Designed by SBG Partners for Culinary Brands)

Air-bubble packaging is now the main option on offer for fragile or sensitive goods. These materials are available in a variety of colours - from carbon black through to pink, such as the Anti-Static AirCap® from Sealed Air Ltd.

Retortable pouches

Retortable pouches can be heated or boiled in water to warm foods and medical supplies and tend to be used to a large degree in medical institutions and by the military, for food and medical supplies. Already there are simple boil-in-the-bag rices and sauces, many of which are also available to the general public. Some gourmet foods have started appearing in retortable pouches, too, within highly-decorated cartons, usually bearing four-colour photographs of the enticing meals in tasteful settings.

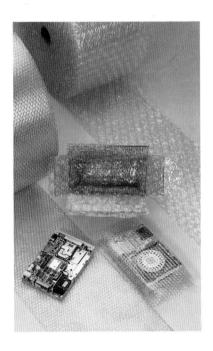

Water-soluble pouches

A potentially stunning design development by Ciba Geigy, Du Pont and others is the dissolving plastics sachet. This can contain precise amounts of agrochemicals, for example, saving farmers a great deal of time: no longer do they have to measure sometimes quite dangerous chemicals in small measuring jugs. The farmer simply throws the water-soluble sachet in a large vat of water, where it dissolves. This form of pack could be used to dispense garden chemicals and dyestuffs, or other materials for general household use.

Antistatic packaging

Other developments the designer can utilize include the antistatic pouch for electronic or electrical components such as computer chips. Charges as low as 50 volts can destroy or degrade these sensitive components and charges of up to 4,000 volts can be generated simply by opening a standard plastics bag, or unpeeling packaging tapes. In addition, 30,000 volts can be generated by walking across a carpet. Static is therefore a serious problem. One major European electronics manufacturer reported the loss of thousands of dollars worth of components transported by rail because of static electricity generated in the bogies and conducted up through the floor of the carriage.

Most designers' interest in antistatic packaging arises through trying to satisfy the functional needs of packaging perhaps using conventional materials at first to protect items including television sets, video cassette recorders and washing machines against transit shocks and vibration. The materials used in transit packaging generally include general purpose polyethylenes and so-called free-flow polystyrenes as well as the more traditional corner fittings.

US research into electrostatic discharge (ESD) showed that new materials were needed to prevent static building up and earthing through the increasing number of sensitive electronic components in today's electrical and electronic goods.

Air-bubble cushioning

One of the first materials used for packaging against ESD was the pink air-bubble cushioning materials produced by companies such as the US Sealed Air to reduce the effects of electrostatic discharge as well as to absorb the usual mechanical impacts. Other manufacturers soon followed suit, introducing their own antistatic bubble materials. These materials are light and easy to cut, although they resist tearing. Being transparent, they allow users to see the contents, which reduces the need to open bags and inspect contents, which could lead to static build-up. Another important characteristic is that the material is heat-sealable.

Of course, the materials also have disadvantages. Printing on bubble packs is very impractical and some of the antistatic coatings actually rub off, weakening their effect with time. A stronger material evolved by designers to combat ESD is the black graphite or carbon-loaded bag, which is generally more conductive than a bubble bag. Usually the carbon fibre layer is sandwiched between two layers of film or foam, which can be printed. The bags are, however, opaque and carbon can rub off, contaminating, especially, sensitive items.

"Cosmetic" bubble cushioning

Because of the American lead in electronics and electrical packaging in general, the US developers have long been aware of the need to protect their goods from ESD. However, UK designers have been slow to utilize antistatic materials in pack designs. One of the grounds for the resistance has, it seems, been cosmetic, suggests a major UK supplier, because of the limited range of colours available: "However, developments are under way for marketing a whole new range of attractive and commercially-acceptable colours to meet most requirements."

Sacks

Many kinds of polyethylene (PE) have been used to make industrial and retail sacks for a variety of agricultural and horticultural products. Their ability to keep out moisture and to resist chemical attack makes them the best choice for many chemical fertilizers and insecticides, as well as for animal feeds. For the industrial packaging designer, there is no better way to build up an understanding of these materials than by working with them.

There have been problems with sealing the necks of some sacks; the sealed area is usually the weakest portion of the pack. One design solution is to strengthen this area with an extra layer of polypropylene, or some other semi-rigid plastics.

Printing on sacks

Sacks made of polyethylene may be printed with up to four colours on both sides, often with special non-slip inks, for buyers who need to stack sacks in a retail store or in a farm shed, for instance. Lately, plastics sacks manufacturers have used up to six colours and improved their graphic designs tremendously. Part of the reason is that plastics sacks are finding more and more outlets in retail stores, as well as the traditional, bulk-buying industrial sectors. ICI has frequently stated that mail order is the largest potential market for plastics sacks, but at present this is still dominated by cheaper paper sacks. These have a homespun look that people like, according to ICI.

Re-sealable sacks

With more people purchasing by mail order, often in small quantities, though regularly, most plastics sack makers believe that small and medium-sized re-sealable plastics sacks will become an attractive proposition, in Europe as well as in the USA, where mail order is well-established in all retail sectors. Already, resealable plastics sacks are used for postal sacks by couriers DHL and Federal Express. Another indication that plastics sacks will become more common is that the cost of paper sacks is steadily increasing, in Europe at least.

Design developments

Other product areas where the plastics sacks could offer distinct design advantages include the many non-agricultural areas, such as pre-packed coal and aggregates - sand and cement - for the construction industry, where product competition has been increasing in recent years. Cement, for instance, has traditionally been packed in paper because it is packed while hot. Most plastics sacks, however, cannot withstand the temperatures involved. Now, manufacturers including ICI have developed four and six colour heavy duty sacks for cement and other hot-fill materials, although these are very expensive forms of packaging.

Another important trend shown by sack designers is "down-gauging", a reduction in thickness of the material used because the films are becoming stronger. A slight problem with this is that thinner sacks lose their stiffness, which makes it difficult to seal the floppy, open ends. ICI is now working with industrial designers, though, exploring the possibilities of ribbed or embossed sacks to ease the problem.

Perhaps the most important trend as far as the graphic designer is concerned is the ability of most manufacturers to print in six colours rather than four. In the past, sacks have been intrinsically dull, but as retail outlets now sell more sacks direct to the public than ever before, sack design takes on an increasingly important role in product marketing. This increasing retail strength is typified in the UK by the boom in garden and DIY centres as well as in retail outlets of the builders suppliers.

Other developments that designers are investigating around the world include breathable plastics sacks for packing powders or vegetables, such as potatoes, and there is an accelerating interest in the packing of goods at high-temperatures.

It will not be too long before making products tamper evident will be a legal necessity. Shrinkwrapping the necks of bottles will then become commonplace. Note on this olive oil container, designed by the Michael Peters Group, the red tear tape that enables the consumer to remove the seal easily. Beneath the seal is a roll-on metal closure with neck-ring, which also acts as tamper evidence.

Decorating horticultural products such as peat, potting composts and gravel would once have been thought wasteful. Today colour and graphics are used to good effect on materials ranging from high-density polyethylenes to special, sunlight-resistant coextrusions and laminations.

Tamper-evident packaging

There has been increasing concern in recent years over cases of deliberate contamination of foods and pharmaceuticals, leading to a demand for "tamper-evident" packaging (no packaging can be guaranteed "tamper-proof"). One solution is to use plastics such as PVC or viscose to seal caps and closures by shrinking a sleeve over them. Though this provides security, it can also be prove difficult to remove for some users, especially the elderly. Manufacturers tend now to incorporate perforations, a tear tape or a tab to enable the seal to be opened.

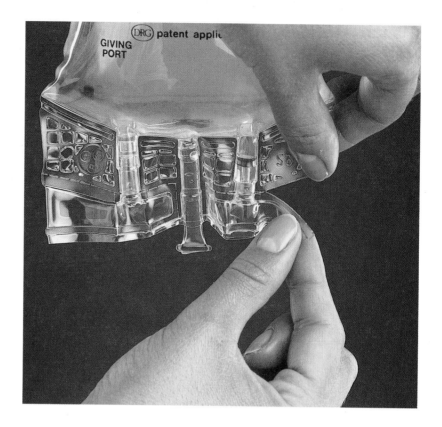

From applications ranging from foods to sophisticated electronics and engine components, plastics films offer one of the most flexible design materials of today. It is no surprise, then, that materials from the plastics world are essential in the packaging of essential sterile medical fluids, especially for use in intravenous therapy.

There are three main uses for specialized plastics bags and pouches: intravenous therapy, including drug-dose applications; irrigation, and dialysis.

The UK's DRG Flexpak has acquired the rights worldwide to a series of designs for polyvinyl chloride pouches and bags for clinical use in these areas. The bags - ranging from 50ml to 5 litres - incorporate a variety of safety features to ensure the protection of the various products, as well as clinical safety to the doctor or nurse using the packs.

The essence of all the designs is that the bags bring a simple, 'touch-free procedure', with little risk of contamination, according to DRG Flexpak. 'The patented internal ports for drug administration are achieved by a unique manufacturing process which enables sterility to be maintained after the sterilization process at temperatures up to 121 Celsius.' This is achieved because access to the outlet ports is only possible via a patented tear-off feature.

The success of the pack can be assessed from its track record - after almost a decade in use, the bag has been adopted by designers of pharmaceutical fluid containers around the world.

For transport, the bags have been designed to be supplied flat and sealed against contamination. For safety purposes, quality control data is embossed on the bag, and the batch and lot number data are printed on the bags when they are about to be filled. The product-formulation data should be printed using the hot-foil method, suggests DRG, as the printing plate bearing the numbers can easily be changed.

Left: when the nurse has checked the bag and removed the overwrap, the drug additive port can be revealed by removing the tear-off feature. The drug can then be added.

The Flexpack mini-bag shown in use right, with a drug transfer facility which has been designed to ensure 'solution' of dry powder or lyophilized drugs and transfer to the mini-bag. A snap-off device is available to plug the administration port after drug transfer, preventing the addition of further drugs.

Below: a hot foil printing machine.

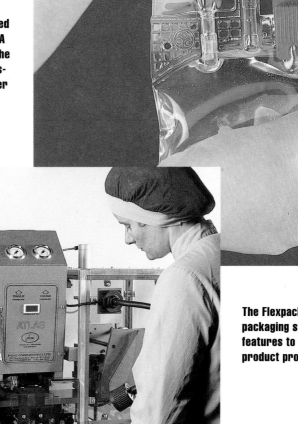

The Flexpack, below, is part of a packaging system with many design features to ensure the ultimate in product protection and clinical safety.

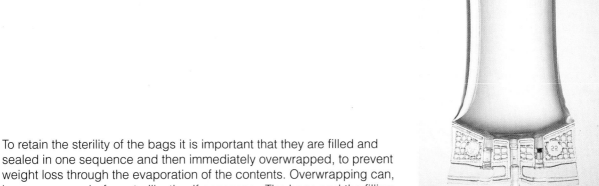

To retain the sterility of the bags it is important that they are filled and sealed in one sequence and then immediately overwrapped, to prevent weight loss through the evaporation of the contents. Overwrapping can, however, occur before sterilization if necessary. The bags and the filling systems comply with international pharmaceutical standards.

*2.*3

Foils and speciality films

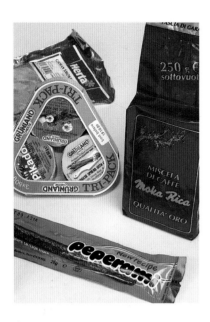

It is hard to imagine that, once, pure aluminum was considered to be a precious metal. However, in the early years of the nineteenth century it was a crown of aluminum that graced the King of Denmark's brow. The pure metal had only recently been purified from its ore. In France, too, the metal ranked highly and Napoleon III used a table service made from aluminum.

It took many years before the extraction process became commercial, but its first use in packaging came - as with so many other materials - from the USA. In the last years before the twentieth century, Ball Brothers began to use aluminum covers for Mason jars. Then, just before World War 1 began in Europe, the first foil wrappers were seen, for chewing gum and Life Savers candy bars.

Foils - on their own, or in combination with other materials - are essential in the packaging of food and pharmaceuticals. They are light, malleable and add value, making them one of the designer's prized possessions.

Aluminum and tin foil

Most of the foils used in packaging today are made from aluminum rather than the expensive tin foils. Aluminum foil is still relatively expensive, but most designers feel that its properties are worth the added cost. Designers should exploit the metal's lightness, strength and durability, as well as its other, more specific properties, such as thermal and electrical conductivity. Important points to remember about aluminum are that it performs well in low temperatures, resisting stress and cracking, and it is not magnetic. This helps to sort aluminum from tinplate cans in the recycling process.

Tin foils tend to be used where their own special chemical properties are needed, but as pure tin becomes increasingly expensive to mine and purify, it seems likely that its importance in packaging will diminish.

World usage of aluminum foil	
Application	%
● flexible packaging	54
● semi-rigid trays	38
● laminations and coextrusions and decorative labels	6
● cap liners and seals	2
Source: industry estimates	

Until the advent of special techniques for coextruding plastics and for laminating them, one to another, many products - notably confectionery and snack foods - have had to suffer second best. Now these products can be preserved better than ever before using materials that can be printed with the most extravagant designs. (Photo MCG Venus Packaging)

What is a foil?

A useful working definition of a foil is that of a rolled section of metal less than 0.15mm in thickness. At its widest it is generally 1.52 metres, though some of the heavier gauges can be 4.06 metres wide. Foils generally vary in thickness by plus or minus 10 per cent, and it is worth bearing in mind that the flatness of foil, often called its "shape", is difficult to control once it is drawn very thin.

Most foils are rarely 100 per cent pure as the process leaves trace elements in the metal. The purest form of aluminum, containing only 0.01 per cent of impurities, is used to make capacitors for the electronics industry. Most commercial forms of aluminum used in packaging are around 99.5 per cent pure, with up to 0.4 per cent of iron and 0.1 per cent of silicon, making up the balance. Increasing the amount of iron in the alloy raises the strength of the foil, which helps in the packaging of tablets or capsules in strip or blister packs, but this may also lead to corrosion. To provide greater stiffness in semi-rigid trays, manganese and copper are often added.

Household foil is now rolled from a special alloy instead of commercial-purity metal and it has subsequently been possible to reduce the thickness of the foil by about 15 per cent in recent years because the newer "aluminums" are much stronger. Semi-rigid containers for pies and frozen foods are also about 20 per cent thinner than they were some years ago, because of the special alloys designed for strength and formability.

Down-gauging

This reduction in thickness or down-gauging is all part of the metal manufacturers' plans to ensure aluminum remains a light, attractive and cost-effective packaging material to compete with the high-performance plastics films, discussed in Chapter 2.1

In the UK, printing and packaging now account for over 50 per cent of the aluminum consumption, compared with 30 per cent in the 1970s, although there are significant differences across Europe. The UK has an above-average consumption of aluminum cans, for instance, but consumes relatively little in the way of foil. In France and Germany, however, aluminum is not generally used for cans but for the more common foil containers. In Germany, even pet food is designed in aluminum foil containers.

Characteristics of foil

Although aluminum foil may seem to be a thin, easily torn material, it is, in fact, one of the best protective media in the packaging field because it is almost impervious to moisture and oxygen. This makes aluminum foil ideal for packing equipment or machinery for the export trade, where corrosion is a major problem. An added bonus is that it is very attractive and can be decorated by the designer in various ways. It can be embossed or brushed - to add texture - or it can be coated with inks to provide a reflective metallic finish.

Another useful characteristic is the deadfold capacity of aluminum foils, to hold creases and to be moulded into almost any shape. However, deadfold foils wrinkle easily and unless the material is being used as decoration manufacturers usually recommend burying the foil inside a container or within its walls to protect it from harm, such as in the walls of fruit juice cartons.

The trend to original pack dispensing in the pharmaceutical trade has lead to a major change in the way drugs are packed. Blister packs - small sheets of rigid PVC bubbles backed with foil - are now widespread. Besides being tamper evident they ensure that if a pack says 100 tablets it contains 100 tablets. Some blister packs have been criticized for being difficult to open - an important factor for the designer to consider if the drugs are intended for the elderly or infirm. (Designed by Packaging Innovation)

Reinforcing foils

Where aluminum is used in slightly thicker forms, or reinforced with dimples or ribbing, it can be extremely effective as a closure or lidding material. The Sealed Safe closure, designed by the Speciality Packaging Group in association with Alcan Aluminum, uses a 60 micron aluminum membrane coated with a special heat-sealing layer. The membrane is sealed tightly to the rigid lip of various containers, including composite paperboard canisters, as well as metal and plastics cans. A pull or ring-tab enables the buyer to peel it back, and the manufacturer's claim that the membrane cannot be resealed. This makes the closure particularly appropriate for tamper-evident applications, such as medical or pharmaceutical packaging. Other potential applications include food and beverages, light industrial oils and hand cleaning grease compounds.

Although aluminum can resist solvents and greases well, resistance to strong acids and alkalis is quite poor, unless it is protected by a wax or lacquer coat. On the other hand, foil will protect a pack's contents from sunlight and will often be used therefore to pack sensitive medical supplies. It is unaffected at temperatures up to 550 Celsius, so surgical blades and syringes can be sterilized within foil pouches where necessary. Pouches can be made for bulky as well as petite items by forming pockets that take the shape of the proposed product.

However, the designer must remember that aluminum is quite weak and tears easily at thin gauges - in technical terms it has a low tensile strength. For this reason, printing in colour and holding registration on unsupported foils is particularly difficult. So it is often worthwhile cultivating companies whose engineers can deal with thin foils. An ability to print thin, unsupported foils, for instance, will enable designers to use it in high-quality goods for the confectionery market - particularly in the foilwrapped figurine and Easter egg field.

Because, in its thinner forms, foil is too weak to use on its own, it is usually combined with kraft paper (see Chapter 2.1), which adds strength and stiffness. However, other, more attractive papers can be used, if appearance is important, as kraft can look dull. The cheapest way of fixing aluminum to paper is to use a solution of sodium silicate, which is widely used in the preparation of cigarette and soap bar wrappers.

Printing on foils

Foils are usually printed using flexography, but, as with papers and plastics film, long runs and high-quality work benefit from a switch to the gravure process. Because the foil surface is shiny, it is often necessary to use a primer and the type of primer will depend on the particular ink or treatment chosen by the designer. Coloured lacquers, for example, can give the appearance of gold, copper or other metallic effects, but these each require a different primer, depending on the sophistication of the design.

While foil-laminated packs enable the most exciting of designs to be printed on them, they do have a disadvantage over the traditional cellulose or polyethylene wrappings in that the product can not be seen unless a separate window is welded in to the design. These snack foods, designed by the Michael Peters Group, for UK retailer Marks & Spencer show a photographic window to compensate.

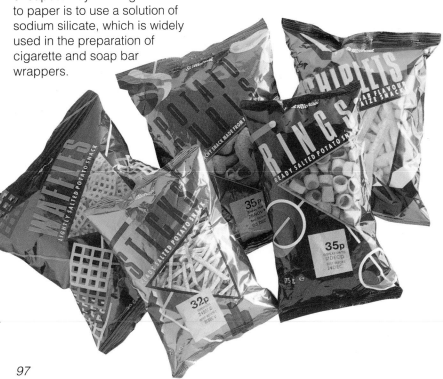

Metallizing plastics

Perhaps the first example of a metallized plastic was the development of tinsel one Christmas in the 1930s. But it is only in the last decade or so that the true potential of metallizing plastics for packaging has been realized. Because metal foils are so impermeable to oxygen and moisture, manufacturers have long sought to bind or laminate them to other materials to pass on these characteristics. A cheaper method of giving a plastics film the characteristics of foil is to coat it with particles of the vapourized metal in a vacuum chamber, in the process known as metallizing.

This range of snacks from KP was launched to break new ground in a large but static market. The foil/oPP lamination bears printing well and the packs have a long shelf life, which is important for these sensitive foods. The cartoons evoke the country of origin and target a certain kind of buyer. (Designed by the Michael Peters Group)

Polyester is the most usual material or substrate for coating. In the United States, Golden Flake Snack Foods chose to leave behind its traditional glassine bag in favour of a reverse-printed oPP, metallized polyester - Melinex 850 from ICI. Besides looking more attractive, the company's designers claimed it also kept the crisps fresher for longer. The switch with single-serve packs was an experiment, says the company's purchasing director, but one that proved so successful that the mid-range bags were switched too. Polyethylene, nylon and polypropylene films are increasingly used.

Metallized plastics films, in comparison, are not often used alone, but are more usually combined with other films for ease of machining. A typical combination is metallized polyester combined with polyethylene, as now used in some ground-coffee and coffee-bean bags. The bags act as a one-way valve, allowing the gases emitted from roasting and grinding to escape gradually without letting air in - the process takes nine months. These new valved bags extend the shelf life of coffee to one year, and are more cost effective as they can be printed directly, unlike the hard vacuum-packed coffee bags that usually require cartons.

Paper can also be metallized, providing it is lacquered to prime the surface. This can be successful on a number of counts. First, it has every appearance of being as attractive as a true foil, but costs much less. For instance, cigarette wrappers formed from laminates of foil and paper cost around 25 per cent more to produce than metallized papers.

These paper/foil laminations allow the printing of half-tones and artwork on the outside, while the foil inside protects the contents from oxidation by the atmosphere. The layout of the facing with the company lozenge and the photograph depicts the variety of the vegetables inside. (Designed by the Michael Peters Group)

Laminations

The basic lamination process involves the combination of two or more materials from separate reels with adhesives. However, there are a number of options open to the designer, depending on the materials that need to be combined. Molten plastics can actually be extruded over a moving sheet or web of paper, though the process is a little more complex and the end result a little more costly. The designer should always consider paper first because it is the most economic substrate and because it adds stiffness to a material.

For the designer, printability is also an important consideration: laminated paper sacks with foil/polyethylene liners can be printed in the normal way with four or six-colour graphics. As well as being decorative, these sacks are well designed technically in comparison with most sacks and can be designed to meet a variety of applications, for example, they can be used for sensitive products such as animal feeds and milk powders.

Before choosing a laminate, designers must first ensure there is no single film capable of doing the required job. Lamination is an expensive business, especially when the process is carried out to the designer's order. If a single film cannot manage the task, the various coatings available should then be considered before lamination, again on cost grounds.

Among the common food applications for laminated films, such as oriented polypropylene oPP/Cellophane laminations, are snack foods. Meats or cheeses need something stronger, such as nylon/PVdC/polyethylene, while oPP/Cellophane laminations are most often used to wrap confectionery.

For a product that prides itself on freshness and taste, to show the product is absolutely necessary to entice the customer to purchase, as here with US snack food, Santitas. (Designed by Apple Design)

Polyester/polythene laminations

Non-food applications for laminated films are also multiplying as designers learn to work with technologists. Polyester/polyethylene laminations, for example, can be used to coat drums against corrosive chemicals. Paper/PE materials, in addition to their use in cigarette and stock cube wrappings, can also be used to wrap analgesic tablets. These and weaker ointments can also be wrapped in Cellophane/PE/foil/PE laminations. Stronger ointments and cough syrups need foil/PVC laminates, often in the form of a container or as pouch linings.

As always in packaging, when choosing laminated materials, it is critical to understand the behavioural characteristics of the product to be packed. If the product is a fresh vegetable, for instance, which gases will it give off? Any proposals for printing or decoration should be discussed with plastics and foil engineers before a lamination is finally selected.

Plastics coatings

Technical developments in papermaking, as well as in plastics, have opened up new possibilities for the designer. The increasing number of microwave ovens, for instance, has stimulated the development of bags and paperboard containers suitable for high temperature cooking. Manufacturers such as American Can, Westvaco, International Paper and John Waddington regularly announce new materials and packs for this market.

Bags of popcorn and poppadums are regularly seen in supermarkets and, like paperboard trays, can be attractively printed. For convenience' sake most microwaveable papers and paperboards can be served at the table, which is an increasing requirement for foodstuffs in the USA. Perhaps the most important part of these packs, though, is the 0.00125 in. (0.03 mm) polyester or acrylic coating that protects the paperboard from soaking up juices and gravies at temperatures of up to 450 degrees Fahrenheit. Higher temperatures may cause the paperboard to smell and taint the food.

Coatings and foods

Significantly, there are restrictions on polyester coatings for foods in the United States and Europe, notably in Germany, as there is some evidence to suggest that an important ingredient of the coating causes tumours in some experimental animals. This does emphasize the need for designers serving international markets to first familiarize themselves with important local legislation before embarking on new projects.

Plastics coatings can also be used with plastics, but if, after research, it is decided that a lamination is absolutely necessary for the required thickness, the number of plies should be kept to a minimum. Another cost-cutting exercise designers should employ is to use the lightest possible gauge or thickness for each ply, short of making the material too floppy to work with. In addition, the most inert, most protective material should generally be specified as close as possible to the product within, not as the middle layer in a carton wall, for instance.

Coextrusions

Another speciality film is the coextrusion where two or more melted plastics are blown through a slit die. There are two chief methods to produce coextrusions, also known as coex films: the blown method, where molten plastics are blown through concentric rings to produce tubular materials, and the cast process for producing flat films on a reel. Almost 70 per cent of coextruded films use polyethylene or polypropylene as the main substrate; the other films in general use include nylon, PVdC and styrene.

Adhesion between web and coex can be a problem with some films, but a thin layer of another plastic, which sticks to both, can often be interpolated. For short runs coextrusion is expensive because of the amount of waste material generated by the process, but for longer runs it can prove extremely cost effective.

Design considerations

As far as the designer is concerned the details of the two coextrusion processes are relatively unimportant but the resultant films can have very different properties. Blown films tend to be used for the more heavy-duty tasks such as horticultural, refuse and postal sacks, while cast films are more often used for quite sensitive materials: packing frozen foods and wines, for example, in bag-in-box containers. As always in packaging, there is some overlap, but the distinction is fairly acceptable.

There is one major disadvantage of coextrusion, however: foils cannot be incorporated within them. And for some designers, bearing in mind foil's extremely useful qualities, that may well be the deciding factor.

The modern pack designer has a major part to play in utilizing both old and new materials in these innovative packs. Indeed, because of their growing exposure to new packaging forms, modern consumers expect nothing less.

These part-baked rolls and baguettes can actually be cooked in these plastic bags. The film - a multilayer polycarbonate from Metal Closures Venus Packaging - resists the high temperatures necessary for baking the loaves and even protects the bread from burning. It takes printing well and the design suggests French origins with colour, imagery and typefaces.

*3.*1

Board-based containers

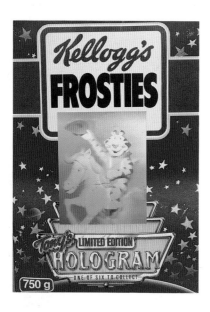

The carton is most definitely a US development and, like so many of the world's great inventions, it came about by accident. In Brooklyn in the 1870s a printer and paper-bag maker Robert Gair was printing a series of seed bags. During the print run a metal rule used to crease the bags worked itself upwards a few millimetres and actually cut the bag. Gair immediately realized the advantages that cutting and creasing board in one operation could bring and developed a whole series of carton-making procedures. He is alleged to have held more carton patents than any other individual or organization. Soon, however, other people began to devise cartons and today most of the techniques of carton-making bear the names of American inventors.

The design for the first carton product, Kellogg's cornflakes, has changed little over the last half century and confounds the argument to redesign and relaunch regularly. However, as shown above, the most up-to-date features have been used to target customer segments as with this hologram on the children's Frosties cereal.

The Kellogg contribution

One of the first users of the carton were the Kellogg brothers, who first devised their famous flaked breakfast cereal at the Battle Creek Sanatorium. The cereal was originally developed as a health food, but the marketing intuition and flair of W.K. Kellogg turned it into a healthy food for everyone.

Kellogg soon arranged nation-wide sampling sessions, hiring vans bearing the Kellogg's signature to take the cereal to the people. He devised grocery displays and gave away free sample packs. In one of the first examples of product branding, he took the biggest ever billboard in Times Square and ran his signature diagonally across the largest section of the sign. The pack still bears his name today. "If it doesn't say Kellogg's on the packet, it isn't Kellogg's in the packet," now becomes a slogan against the own-label cereal manufacturers.

In 1915, to take further advantage of the health trend, Kellogg launched its 40% Bran Flakes and followed this in 1916 with All Bran. Both brands are still on sale today, as is Rice Crispies, developed in the 1930s to appeal directly to children. Much the same approach is shown today in on-pack decoration and television advertisements for other products such as Weetabix.

Carton inners

In the 1920s and 1930s a series of patented packs were developed by Kellogg that were rapidly taken up by other manufacturers. The first was the waxed, heat-sealed bag known as Waxtite, which was initially sealed around the outside of the box to keep the cereal fresh merely in transit, but then it was transferred to the inside of the box, freeing the surface of the carton for printing and advertising. The bag is still there today, but more often than not it is made from plastics, which afford the cereal an even longer shelf life.

Of course many more products appear in cartons today than in the 1930s and 1940s. Leading the way are cosmetics and toiletry products, foods, proprietary pharmaceuticals and tobacco products. The carton is convenient, clean and compact, but there are still improvements to be made: new and improved methods of coating, folding, opening and closing. These are the province of the packaging designer.

In this range of packs for Freedom hygiene products, designed by Studio Kreuser of West Germany, board has been used to liven up a traditionally dull product. The folding carton using the cutout window is an interesting departure, but the hexagonal dispenser of handbag-sized packs is a classic in the making; while the discreet graphics and attractive colouring make it a container that will not look out of place in the home.

To increase the sales of tissues, tissue manufacturers had to encourage people to keep tissue boxes out in the open, ready for use at the slightest need. To do that they had to provide designs to suit every taste, for every room in the house, for the bathroom and for the office. Shown below is one of the many designs for Kimberley Clark by Studio Kreuser.

Designing cartons

There is an arbitrary limit in the packaging industry which suggests that a carton is any holding receptacle made from board between 250 microns (0.010 in.) and 1,000 microns (0.040 in.) thick. Outside these limits are paper and paperboard - below 250 microns - and corrugated fibreboard cases or rigid containers - thicker than 1,000 microns. But this is where the definitions stop, for cartons can be of almost any shape and construction. Depending on the finish, the quality of board the designer chooses and the ability of the printer, cartons can be made into a delicate gift pack for jewellery and cosmetics or a simple container for screws or nails. The traditional method of designing the carton follows much the same pattern, though, whatever the product:

- a carton must contain the product, allowing it to be transported and dispensed, if necessary, with ease
- it must protect the contents from breakage, spoilage, moisture uptake or loss and leakage
- it must advertise the product to the consumer
- it must sell the product to the consumer

Packing individual bottles of ink may seem a waste of money, but the attractive designs tempted retailers to devote more shelf space to the range and appealed to ordinary consumers as well as professionals. The designs, by the Michael Peters Group, soon came to be regarded as "collector's items" so it became natural for consumers to covet the set.

Quality and construction

Designers considering a carton as a container for a product must establish what hazards the product is likely to face to enable the right quality of board to be chosen and the right construction - some boxes are inherently stronger than others. If the contents are waxy or oily, for instance, a waxed or plastics-coated board should be used.

As with all packaging design, the carton must attract the customer's attention and hold their interest. The outside surface of the carton is therefore as important, if not more so, than the inside: it is the only advertising medium the products may have in the retail environment. If the pack advertises effectively, it must then sell the product. It must tell the potential buyer what the product is and how much is inside. It must also bear the brand name and explain why this particular brand is better than other brands, or even own-label products of the same kind. If necessary, the carton may also have to explain how to use the product. Even well-established brands need this if they are to attract new users.

The carton as visual reminder

All these facets of design must tie together and give the customer a sound reason for buying. But the carton's function does not stop suddenly, at the checkout. Wherever the product's final destination - at home, in the garden or in the office, for example - it must provide a strong visual reminder that the product is a good one and worth buying again.

This all-important, buy-again factor is often forgotten by packaging designers. There has been a trend in recent years, for example, to devise tissue cartons with a rip-out central panels, which is also the only part of the pack bearing the manufacturer's name. This presumably is meant to encourage the buyer to leave pretty cartons about the house. Bedroom and living-room designs regularly appear. Designers even use cartoon characters on boxes for children's bedrooms. Having ripped out the brand name though, some consumers do not recall whose pack it was. In the absence of detailed research, this calls for clear and sometimes lateral thinking on the designer's part to determine exactly how a product will be used by the purchaser.

Graphic design

Having established the design criteria for the proposed carton, the designer then moves on to consider the graphic design of the pack. This is important at an early stage, as it often affects the type of board and its finish. Four-colour photographs, for instance, require a different treatment from two-colour or monochrome line drawings. One effective design feature is to display the product itself through a window in the pack. The visibility of the product, especially if of high value, can lend tremendous weight to the sales message.

After the design has been proposed, the structural considerations need to be taken into account. There is a comprehensive list of requirements which all carton makers will be glad to help with, but there are some basic factors to consider:

Considering all these aspects of pack design should give the designer ample information to put design proposals to the customer. The proposals should also include the type and thickness or caliper of the selected board and the number of colours to be used in the design. The proposals should also go into reasonable financial details, outlining how much preparing photolitho positives and printed proofs will cost, for example. It is worth listing any additional processes separately that might be necessary: these might include a high gloss varnish process, metal foil blocking or bronzing, together with embossing or coating with latex for protection.

- are any fitments or dividers necessary to hold the product within the carton? Should these be moulded from foam or sponge rubber? Should there be some form of cushioning? Can the fitment be another piece of board?

- is an outer sleeve, carton or label required?

- is the product seasonal?

- is there more than one product in the range? If so, the design has to work at all the different sizes.

- how will the cartons be filled and sealed?

- how will the products be stored?

- how will the packs be displayed? Are they self-standing or do they need a secondary carton for support?

This range of fresh soups (top) from the New Covent Garden Soup Company uses the traditional gable-topped carton reserved for other fresh liquids, such as milk. (Designed by Brand New)

Packing heavy glass spirits bottles in brightly coloured cartons is a marketing move clearly intended for the gift market. Above, the cut-out letters reveal a picture of the Beefeaters of the Tower of London after which the gin is named. (Designed by Mittleman/Robinson.)

Types of board

It is clearly imposssible for a board manufacturer to hold every type of board in stock. However, a wide range of boards is usually held by most carton makers. In addition, there are ranges of coatings that can alter board characteristics, such as water or wax proofing. In recent years there has been a great demand for foil as a decorative medium and many millions of cosmetic cartons are now made from board laminated with various foils.

Every board type must meet certain basic requirements. The outer surface must give good results when printed using the offset litho printing process - the inks must adhere or key well to the surface. The board should also be receptive to adhesive. The board must also handle well and glue quickly on high-speed gluing machines. In addition, the board must crease well, without cracking or fracturing, and it must form the required carton shape on automatic packaging machines without collapsing.

Multi-ply boards

The multi-ply boards known as duplex boards are made from wood pulp surfaced with a bleached kraft liner. This is well-suited for printing solids and half-tones (gloss), which makes them effective for cigarette packs and pharmaceuticals, as well as various foodstuffs. Solid white boards are formed from bleached wood pulp and are white throughout. They are available in coated or uncoated forms and have good printability. They are generally used to project a high-quality image, in cosmetic packaging, for example. Chipboards, in comparison, are manufactured from wood pulp with recycled pulps. These cartons have a grey back and are well-suited for line printing. These cartons tend to be used for disposable outers, as in oven-ready meals, or as fittings and dividers in larger cartons.

Finishes

Duplex and solid white boards can be cast-coated to give a quality finish. This makes them heavier and the smoother coat gives a high gloss after varnishing. The process is expensive, though, and these cartons tend only to be used for ultra-high quality products such as cosmetics and jewellery. For packs that really want to command attention, designers are trying foil-lined boards. The process of laminating can, however, be very expensive, though the effects of matt or gloss gold and silver can be very attractive. Boards may be laminated or coextruded with plastics to provide grease or oil resistance. These cartons tend to be used for foods and waxes.

In 1988 the packaging for these Harley Davidson cigarettes, designed by Mittleman/Robinson, won a major US award for packaging design. The combination of black and gold is set off by the subtle red line.

A number of special effects can be achieved with boards - perhaps the most expensive being foil blocking and bronzing. In the blocking process a brass or copper blocking die-stamps foil from polyester film under high pressure and temperature. The blocked image can be flat or embossed and it can have a matt or a gloss finish.

In bronzing, a special varnish is applied to the requisite area on the carton and a fine metallic powder dusted on as it passes through the bronzing machine.

To meet the growing demand for alternatives to the expensive foil-lined boards and new metallic inks, improved standards of gravure printing have been achieved in recent years, far exceeding the capabilities of offset lithography. In food cartons, for instance, high quality gravure inks that do not smell reduce the chance of residual odours penetrating the foods. This was a particular problem with some confectionery cartons in the past.

The familiar embossing of letters, logos or other shapes is achieved by placing the board between male and female dies and then applying pressure. This is sometimes carried out simultaneously with cutting and creasing. Other increasingly familiar sights are the varnished boards, carrying a variety of finishes from matt through to high gloss. Varnishing can help a carton resist rubbing, alkalis and the cold temperatures in freezers, among other things.

Applications

There has been a distinct trend, in high quality design, towards metallized, laminated and embossed boards, for example in the packaging of chocolates, such as the chocolate pyramid products from UK Terry's of York. There are many smaller paper and board manufacturers also enjoying success with high quality product designs. Breakfast cereal manufacturers, for instance, are benefitting from the consumer's concern with health and nutrition: premium high fibre cereals are in. US and UK brand leader Kellogg's, for example, still relies heavily on bag-in-carton packaging, but an increasing number of minor brands and own-label suppliers use flexible packaging with no outer container.

Though biscuit products are still packed in flexible films, increasingly oPP, there is an interesting tendency in the premium products sector to return to the carton outer. These are printed in up to six colours, highlighting the trend towards quality products, even in the own-label area. Most crackers and water biscuits, an important sector of the UK biscuits market, still use a corrugated inner to protect the delicate biscuits.

Frozen foods

While frozen food and ice cream continue to be launched, increasing the market size by around 5% each year, plastics containers are increasingly used for bulk purchases. Cartons are much preferred for sophisticated prepared meals, such as curries, and other foods in sauces. Important UK packaging suppliers in this area include Field Packaging, DRG, Taylowe, Metal Box and Robinson and Sons.

The ready-meals market has for many years used cartons, because they are the only medium that can reproduce the half-tones that will sell the final product. With the increasing sales of microwave foods the safety of the carton in this market seems assured. But design competition is increasing and so the innovative use of type and illustration, as in the range of Caribbean Classics from Goya designed by Image shown here, will continue to increase.

Ovenable trays

Another increasingly important area for modern designers is the convenience food market which requires ovenable board trays. Europe's microwave population is now almost as large as the USA's and is certainly strong enough to support mass production of these new packaging forms. A significant technological development here has been the introduction of so-called dielectric boards. These consist of metallized films laminated to board and they accelerate the browning of certain meats in microwave ovens, which has made the use of the microwave even more attractive to the consumer.

Another innovation in this sector was Diotite from Metal Box. This ovenable tray system uses polyester-coated board blanks for high-quality cuts of meat in the convenience sector. The Canadian group Lawson-Mardon and Reedpack's Field Packaging have similar systems and five important customers were soon snapped up between them: Dalepak, McCains, Imperial Bakeries, Moray Seafoods and the Co-operative Society.

Spirally-wound containers

A packaging form that has had some small success since its re-incarnation a few years ago is the spirally-wound paperboard container, commonly found as the inner tube to products such as toilet paper and kitchen papers and foils. Now these are appearing as containers in their own right, often aluminum-lined. John Waddington's Ultrakan is perhaps the best known in Europe, although two others - Akerlund & Rausing's Cekacan and Bonar's Canshield - are also well-established. Cekacan is also licenced in West Germany, by Europa Carton. Ultrakan's biggest success has been in yoghurt packaging, showing the effectiveness of the medium at containing liquids as well as dry goods. Cekacan's first contract was a powdered milk product from St Ivel.

Drinks cartons

One of the consistently innovative design sectors in paper and board packaging has been drinks cartons. The market for fruit juices is another beneficiary of the new trends in healthy eating. The one litre pack from companies such as the Scandinavians Tetra Pak and Elopak, together with the UK-German venture Bowater PKL, has by far the greatest sales volume. But the 250ml single portion pack in some sectors almost matches it. Bowater PKL has also introduced a two-litre pack for pure juices, but whether or not this will be taken up on a large scale by the retailers remains to be seen. An interesting development incorporated into the carton is a cutout to enable the user to provide an air hole to make pouring easier.

Fresh milks and yoghurt products also continue to tax designers to the limit, particularly in the Ex-Cell-O and Elopak gabletop carton. One of the first milk cartons was in fact devised by the German company PKL, in the 1930s. But perhaps the most interesting developments are coming in aseptic, long-life cartoning, which has helped to make possible the carton-packed liquid with chunks or particulates, such as soup, traditionally a canned product.

The traditional gable-top carton from Elopak in Europe, licenced from Ex-Cell-O in the US, carries most of Europe's fresh milk and juices. The carton is relatively expensive to produce, but its convenience easily outweighs the cost. Printing is usually restricted to simple line drawings and two or three basic colours, as seen here on this carton from the Mackie dairy.

Aseptic packaging

Aseptic packaging refers to the conditions under which the product is packed, i.e., the independent sterilization of product and pack, followed by the filling and sealing processes. The product - a soup or yoghurt, for example - is usually sterilized by one of a number of high-temperature, short-time methods. In milks and dairy packaging this is usually the ultra heat treatment process (UHT), but juices fare better at lower temperatures. The pack itself is sterilized separately, sometimes by exposing it to a low-level radioactive source, but more often by exposing it to steam or the bleach, hydrogen peroxide.

Cartons such as these below from the Anglo-German company Bowater PKL and similar cartons from the Scandinavian Tetra-Pak are most often used for ultra-high temperature treated liquids such as fruit juices and soups. These need no refrigeration and so afford long life to their contents.

The prime objective of aseptic cartoning is to prevent the spoilage of the food or drink product by microorganisms, but an added benefit is the retention of the product's flavour. Arguably, cartoned products taste better than canned products, which are processed for longer periods. Besides treating its contents "better", from the consumer's point of view, the manufacturer can also benefit from aseptic cartoning. Carton makers claim freight savings of up to 75 per cent have been realized by some retailers. Furthermore, the cartons need not be refrigerated, thus saving on electrical costs. The rectangular shape also saves expensive shelf space and offers a wide, easily printable face for well-designed graphics.

Maggi (right) has introduced a soup with croutons, the problems of packaging particulates having now been overcome, and Studio Kreuser of West Germany have produced a design with bright refreshing colours and photographs of the soup ready to serve.

Packaging of particulates

In 1985, perhaps the most significant carton announcement to date was made, from the Bowater-PKL stable: particulate soups were launched in the Combicloc carton for Crosse and Blackwell's Four Seasons range of soups. The flash cooking process enabled the Nestlé subsidiary to develop delicately flavoured soups containing garnish, chunks or lumps up to 25mm square. The use of new materials - board laminated with plastic and/or foil (see Chapter 2.3) - made it possible for the carton to challenge the metal can as a preserving container.

Market research had showed that consumers would probably like the pack and that the supplier could expect to charge a high premium with little difficulty. The 400ml cartons were printed with the high quality rotogravure process to enhance the feeling of quality, and the product's shelf life is estimated at between nine and 12 months.

Bag-in-box packaging has perhaps the greatest potential of any recent board-based liquid container. Its capabilities are appreciated in the wines and mixer syrups market, but elsewhere its use is restricted. This pack from Stowell's of Chelsea shows just how effective the large rectangular pack can be in promoting traditional wine values with economy of price.

The particulate filling ability has enabled Nestlé to launch its redesigned Maggi soup range in cartons for the European market. Another example of a highly viscous and chunky soup is Parmalat's use of the Combicloc carton for its Pais range, on sale in Spain and Italy.

Although one of the most advanced in the field of particulate packing, Bowater-PKL is not the only company in the marketplace. Metal Box and Scandinavian Elopak are working on Odin, a joint venture to bring particulate technology to a wider audience. The research programme is long and detailed, but the companies expect to have a commercial launch on the shelves in the mid-1990s.

Bag-in-box packaging

One of the most significant aspects of the world's growing wine market, especially for retailers and distributors, has been the evolution of new packaging forms. Since the beginning of the 1980s, three major packs have been ousting high-quality glass bottles from supermarket wine shelves: the one litre carton discussed earlier, the drinks can (see Chapter 3.4) and the three-litre bag-in-box or cask. The number of wine boxes sold grows at about 25 per cent each year and seems to be particularly successful in the UK. The German market seems still to be favouring the traditional glass bottle and resisting attempts by German designers and manufacturers to promote the cask.

Wine is not the only liquid favoured by the bag-in-box, though. The Australian-designed plastic and foil bag, irradiated to make it aseptic and free from infection, together with a selection of valve or tap delivery systems supported in a rigid carton, makes the pack an ideal container for sensitive fluids such as blood serum products in hospitals.

However, it is as a wine and sometimes beer or cider pack that this particular container excels. Winebox market leaders in the UK include Stowell's of Chelsea and Colman's of Norwich. Both companies use graphics to disguise the relative newness of the box as a wine pack and the designs usually give an upmarket, natural feel.

Quality

There are, however, questions about the quality of the winebox as a container. In a survey conducted by the UK's Consumers Association, publishers of the prestigious *Which?* magazine, 29 wine boxes from around the world were opened and tested for taste a week later. Only nine of the boxes were found to be "good", three were "bearable", one was "undrinkable" and five were so bad they received the *Which? Wine* "Black Tap". Most significantly for designers considering the wine box as a quality container, the testing panel concluded that boxing wines definitely affected the taste of wine over periods of more than a few days.

There is obviously a need to improve the image of the bag-in-box for drinkers and retailers, and here the designer can help. Better bag materials will continue to provide opportunities for better performance, lower cost or new applications. In addition, the design of new taps, spouts and closures could help improve shelf life and make the pack even more convenient for the consumer.

The brown box is no longer a ubiquitous feature of the shopping trip. In its place are colourful litho-laminated board containers, such as these bedding containers from the UK Next retail chain, designed by David Davies Associates. Besides offering a larger, more effective area for selling the product, the cases are - for the retailer - easier to store and distribute and - for the buyer - easier to carry home.

Secondary packaging

An exciting development for packaging designers is the increasing treatment of previously utilitarian secondary packaging - the ubiquitous brown box - as a marketing medium, by using preprinted boards for transit packaging. Now almost a fifth of all corrugated cases in Western Europe are printed in more than two colours and the proportion is increasing each year. The proportion is thought to be even higher in the US, but statistics are difficult to collate. The main printing methods used are litho-lamination and preprinting for high-quality four and six colour work, while flexography seems acceptable for the bulk of designs at this stage.

The reason for the move away from the brown box is the increasing consciousness of retailers of the importance of their on-shelf image and the evolution of the transit case into a primary pack with sales potential. The higher the quality of the pack, the higher the perceived quality, with resulting increased profit margins.

Export needs have played a key role, too, in expanding the graphic requirements of transit packaging. US agents, for example, make it very plain to Scottish distillers just how important the outer case has to be. The technique can be expensive, but it does compare well with the cost of labelled cases.

Applications

A prime candidate for the preprint treatment are products already using white-topped or clay-coated boards, and cases using labels to define the contents. Designers should not underestimate the impact of preprint. Domestic appliance producers such as Braun and Kenwood have switched to preprint with attractive designs in four colours and varnished, illustrating the products and their uses in the home.

Solidboard cases

Marketing managers within the packaging industry are adamant that although the solid fibreboard case makers have suffered a decline over the years, the slope has levelled out and may, in fact, be on the increase again thanks to exciting new pack designs. This is particularly true in the fresh food industry, where solid cases are pre-eminent for packaging for refrigeration, one of the most advanced forms of distribution. The material is also used widely in the design of packaging for detergents, although the success of the new plastics bottles for products such as Lever Brothers' Wisk have posed a grave threat.

In the early 1970s there was no better barometer for the state of the packaging scene than the designs of and the market for solid fibreboard packaging. The market began to pick up in the mid- to late 1970s, as horticultural applications grew. Here, good design successes were achieved by twinning solidboard with polyethylene linings to help add waterproofing. Another major growth factor was the booming supermarket trade in vacuum-packed cuts of boned meats and poultry, and the growth in demand for eggs. Much of today's poultry produce, however, is packed in polystyrene trays.

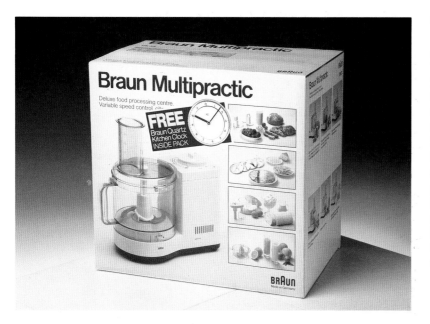

One of the leaders of the electrical appliance business, Braun, has chosen to use high-quality reprographics on pre-printed board-based cases. The sans-serif typeface and the use of a sequence of inserts soon became a house style. Inside, untreated boards act as liners and protect the appliance in transit.

Corrugations strengthen board tremendously and over the years designers and manufacturers have agreed on specifications for the various boards that can be produced. The base corrugations used today are A, B, C, and E flute, as in these double-wall corrugations shown here, offering about 36, 51, 42 and 96 flutes per foot respectively. Tri-wall boards can also be manufactured.

Conclusion

All in all, board remains one of the most flexible packaging materials ever developed. The unsurpassed printability of the surface enables litholaminated B and E-flute cartons, for example, to sport five and six colour halftone illustrations of the highest quality. Its consistency in performance and high stiffness-to-cost ratio means it is a technically-excellent material. And - of increasing importance in the closing decade of the twentieth century - it is totally recyclable.

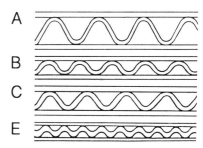

Boxes or cartons are one of the oldest forms of packaging, which means that outside the modernistic food and pharmaceutical packaging areas they tend to be used quite plainly for traditional products, such as shoes or DIY products. This also makes them one of the most difficult packaging media to use in a different and exciting way.

When US agency Peterson & Blyth took on the contract for DearFoams fur-lined slippers, the US's best-selling brand of slippers, it developed the idea of nostalgia as the main theme for the product packaging, and aimed to run old etchings or engravings, in colour, on a plain warm-buff background. But where the designers excelled was in the addition of a light-hearted touch, which lifted the pack from a simple gift or straight-forward functional pack to a talking point in itself. Peterson & Blyth slightly altered the illustrations - adding in DearFoams slippers - bringing humour to bear in a powerful and memorable way.

Working with humour is a delicate matter - no two people share exactly the same sense of humour, but by studying the designs and ideas that Peterson & Blyth's designers assessed - and rejected - during the project, one can soon see the simplicity and impact of the final packages and how these outweigh the other ideas easily.

Another problem the designers faced was the introduction of a new logo - starkly contrasted with the old flowing script, the strong yet refined serif face added a high-quality touch missing from the older design. The use of the metallic gold on ultramarine imbues the packs with a strength the weaker script - white on royal blue - was unable to bring. The successful logo was chosen from a range of similar designs, but the simplicity of the final choice is beyond question. The lines and shadow effects appear far too fussy in comparison.

Strangely, a comparison of old and new packaging highlights how old-fashioned the previous pack was. The new packs are bright, stylish and - most important - warm in their choice of colours and images. Peterson & Blyth couldn't put a foot wrong.

The finished article, supplied flat and showing the cut-away window to allow the buyer to see and feel the foam slippers within.

Right: the new DearFoams logo - simple, striking and memorable.

Left, above: a comparison of the old and new designs shows, perhaps surprisingly, how modern the new designs are. Bold but subtle colours on stylish engravings, tastefully reproduced.

Left, below: one of the original engravings before touching up.

3.2

Plastics containers

The history of plastics is inextricably intertwined with the game of billiards. The connection arises because a US manufacturer of billiard balls offered a reward of $10,000 during the American Civil War, for an ivory substitute, because of the US shortage of ivory. A New York engineer, John Wesley Hyatt, with his brother Isaiah Smith Hyatt, heard of the offer and spent some years experimenting with the action of camphor on pyroxylin, a material made by pouring nitric acid over cotton. The material the reaction produced could not be moulded, but it could be carved and shaped in the same way as ivory. Isaiah called the material "celluloid" and subsequently the brothers Hyatt started a successful billiard-ball manufacturing company in Albany.

Plastics have turned the packaging world upside down, offering shapes and functions never before dreamed of by designers. This so-called oyster or clamshell blister design by Innovations in Packaging for Duracell torches rapidly became a classic.

Over the following decades a series of plastics was developed, including bakelite, cellulose acetate and - in 1935 - the workhorse of the plastics packaging designer, polyethylene, which was developed in England by ICI. The material was used in new blow-moulding techniques devised just before the Second World War and the new material proved useful in the formation of blow-moulded bottles. After much research and development, a squeezy bottle for Stopette deodorant was introduced in 1947.

Using plastic containers today

Many packaging designers today are finding that the first question in choosing a packaging material is whether or not to use plastics. Now, more plastics are used in packaging than any other material. Containers made from plastics are extremely light and can be moulded into intricate shapes in many different colours. More recently, designers have devised a number of different finishes, to counter the challenge from the glass manufacturers that plastics packs can never have a high quality image.

Another major advantage of plastics packs is that they can be squeezed, enabling product to be forced out of a pack. Perhaps the most exciting example in recent years has been the plastics

The ability to mould plastics containers in almost any shape has lead to the development of innovative products that could not otherwise exist. The angled toilet jet made famous by Toilet Duck is a good example. Here is one of the many variants, from Harpic, shaped to fit the hand and the angle of the neck calculated precisely for ease of filling. (Designed by Packaging Innovation)

ketchup bottle. The plastics container has become a familiar and essential part of household requirements, for toothpastes, cleaning materials, toiletries, foods, and so on, due to the innovation of the designers who work with plastics.

In recent years, the trend towards plastics has accelerated and designers have taken more and more of the goods previously packed in traditional non-plastics materials such as board, tinplate and paper, and put them into plastics. A quick survey of the world's packaging magazines shows the increasing interest in plastics as a packaging medium, and the many ways designers are finding to form and decorate packs. For example, a plastics container which looked like china was introduced for cosmetics creams in 1988 by Bramlage of West Germany.

Materials for plastics containers

There are many different materials used to form plastics containers and - as with the plastics films covered earlier - the contents often define which material should be used. In general, though, the plastics which are most useful for making films - the polyethylene family - also play a major part in the manufacture of plastics containers, such as the low and high-density polyethylenes (LDPE and HDPE). Together with polypropylene (PP) and polyethylene terephthalate (PET), these are relatively cheap plastics for packaging and are easily moulded into shape. Also, they are attractive with a shiny high-quality finish. For the characteristics and properties of these plastics see Tables 1 and 2.

An example of the potential of polypropylene is the US development of a one-piece polypropylene box with hinge for fabric bandages or plasters. Polypropylene has very many of the properties which polyethylene has, but is in general much stronger and it is the bending hinge section that enables the container to be moulded in one-piece. The container is also one of the first irregularly-shaped packs to be successfully printed by offset lithography. During its initial trials and during the first sales periods, the container was moulded in two sizes. It was printed in up to six colours and was cured to protect the printed surface. The pack will not rust in a humid kitchen or bathroom and it is less expensive to produce than its tinplate counterpart. Colgate-Palmolive believes the injection-moulded polypropylene container will spread to be used by other products.

Another important plastic for containers is PVC, which can be used to form tubs, trays and bottles, among other things. Its major applications include squashes, cordials and oils, but it has a drawback in that it can split or crack when dropped. This explains the ribbing and extra strengthening that designers build in to PVC bottles to prevent this. Most PET bottles on the other hand are quite plain, except for decorative embellishments, because they need little re-inforcement, except perhaps at the base.

This range of heavy duty HDPE containers, designed by Packaging Innovation for Fisons, is pigmented to protect the contents from the effects of sunlight and colour coded according to which products they contain. The bottles are moulded with a thin strip of transparent plastics along one edge to allow the gardener to see how full the bottles are.

Production processes

In conjunction with the development of this ever-widening variety of plastics, a large family of production processes has evolved. For thermoplastic materials, those that melt when heated, there is the process of *injection moulding*. This can be expensive, but the price per pack falls if the production runs are long enough. Materials which do not remelt when heated are known as thermosetting plastics and containers can be made with these materials in a process known as *compression moulding*. Another plastics-oriented process is *extrusion,* where softened plastics are pushed through a shaped die to form continuous shapes and films. This technique plays an important part in the *blow-moulding* process, which is often described as a combination of extrusion, injection moulding and blowing to produce thin-walled hollow containers.

Table 1   **Properties of materials for plastics containers**   *Source: British Plastics Federation*

Material	Clarity	Impact strength	Rigidity	Barrier		Resistance to chemicals	Resistance to stress cracks	Softening point (Celsius}
				water vapour	oxygen			
MI PS	transluscent	moderate to good	good	very poor	poor	poor	poor	80-100
crystal PS	clear	poor	very good	very poor	poor	poor	poor	80-100
LDPE	transluscent	very good	poor	good	poor	very good	good, if grade well chosen	75-100
HDPE	transluscent	good	good	very good	moderate	very good	good, if grade well chosen	120-130
PVC	clear-cloudy	poor-good	very good to good	moderate	good	moderate	good	70-80
PP	transluscent	moderate	good	good	poor	very good	very good	145-150
oriented PET	clear	very good	very good	good	good	very good	very good	80-90

Table 2   **Characteristics of materials for plastics containers**   *Source: British Plastics Federation*

Description	Primary materials	Applications	Secondary
bottles (up to 2 litres)	polystyrene (PS) polyvinyl chloride (PVC) low-density polyethylene (LDPE)	pharmaceuticals, toiletries, cosmetics, household products	cartons or corrugated cases
	high-density polyethylene (HDPE)	chemicals	plastics film wrap
	polypropylene (PP)	engine liquids	
	polyethylene terephthalate (PET)	foods and beverages	cartons and corrugated cases
bottles and small jerricans (2-20 litre)	PVC	household products	cartons and corrugated cases
	HDPE	chemicals	pallets or pallet boxes
	High molecular weight HDPE (HMWHDPE)	foodstuffs, small industrial packs	free standing
	PP	pharmaceuticals	plastics film wrap
rectangular and cylindrical drums (20-250 litre)	HDPE, HMWHDPE	industrial or agricultural chemicals	pallets, pallet boxes or free standing
		foods, essential oils, flavourings, pharmaceuticals	pallets, pallet boxes or free standing
open topped drums (30-250 litre)	HDPE, HMWHDPE	foodstuffs, paints, pharmaceuticals, chemicals	free standing

Decorating plastics containers

Besides labelling the plastics container, a number of direct printing techniques can be used. For plastics films on the reel or in sheet form, flexography is the chief and most economic printing method, but shaped containers such as bottles, jars and collapsible tubes require offset or silkscreen printing techniques. The silkscreen process should typically be used where the design suggests an embossed effect, and because the process can be expensive it is usually reserved for high-value products. However, the process is somewhat flawed in that it does not reproduce fine detail and cannot satisfactorily be used for high-resolution illustrations or photography.

Another printing technique the designer could use on more rigid plastics containers is hot stamping, where a high-temperature die is placed against a leaf of gold or silver and pressed firmly onto the container. Some of the lighter plastics and thinner-walled containers cannot withstand this process, however, so designers must always check the physical characteristics of the material before they specify this printing process.

This range of hair-care products from Henara, designed by Michael Peters Group, uses the mouldability of plastics to adopt a new shape, paralleled by the label, away from the market norm. This, the company hoped would differentiate the products from own-brand shampoos and conditioners.

Choosing plastics

Although the large number of plastics that exist may seem daunting to the new designer, very often the choice is limited by a number of factors. First, the packing, distribution and storage operations may suggest certain plastics more than others, for protection against moisture, for instance. A bottle for liquids or powders will need a safe cap or closure. Cost, too, plays an important part and when considered in conjunction with the physical properties of the various plastics a shortlist can soon be drawn up.

The product may be covered by legislation, which may also limit the number or type of materials that can be used. In the US, the Food and Drug Administration is exceedingly powerful in this respect, while in Europe only some directives issued by the European Community are effective on a wide scale.

Pharmaceutical regulations

The FDA regulations that relate to toiletries, cosmetics and even foods are relatively simple. However, ensuring that pharmaceutical packaging meets FDA approval can be incredibly complicated. Suffice it to say that the only FDA-approved plastics in general use for rigid containers are polyethylene (PE) and polystyrene (PS). PVC can be used, but only for certain products and the list of eligible compounds changes regularly. The main reason for this difficulty with plastics is that the pharmaceutical and drug makers must *guarantee* that their products are not affected by the containers. This can be very difficult.

Designing with plastics

With thermoplastics almost any colour of container can be produced, simply by adding pigment to the plastics mix. Colour can be a problem with plastics in the thermosetting process, as only the dark colours seem to be effective. Plastics can of course be wholly transparent - the best transparent materials perhaps being members of the cellulose family, as well as polystyrene and certain acrylic plastics.

Protecting products from the smell or flavour of other products - known as taint - can be difficult, as gases tend to flow freely through most plastics. An increasingly successful pack, though, is the so-called high-barrier plastics bottle, essentially a mixture or coextrusion of polypropylene with ethylene vinyl alcohol (EVA). This prevents the passage of gases, usually carbon dioxide and oxygen, because its walls are made from five or more layers of plastics with adhesive between, each plastic having a different property. One of the most significant products to be packed in high-barrier packs is tomato ketchup, a particularly difficult foodstuff to pack because it contains strong acids and sensitive tomato extract.

In this range of preparations from Vidal Sassoon, a variety of plastics is used, from rigid to soft polyethylenes and incorporating valves of polypropylene. Slimline aluminum aerosols are also included in the range, but because of the solid branding by the designer the different materials are submerged and not obvious to the buyer, who only sees the Sassoon logo or lozenge.

For these antifreeze products, below, and the brake fluid from the Q8 company, designer Wolff Olins has used black plastics containers, labelled in black, to show off the colourful sail-like logo.

Weight considerations

One of the major benefits of choosing plastics as opposed to glass or tinplate cans is its lightness. Substantial savings can be made in transport costs, and the eventual convenience to the consumer should not be underestimated. As there is less and less time for shopping and thus more shopping in bulk, the weight of packs must fall. Already this has been seen with the demise of the glass bottle for soft drinks and beers. Automotive oils and greases are also switching to plastics, and the attempts of the traditional material manufacturers to reduce the weight of their products has only prompted designers of plastics packs to cut the weight of their products, too.

Beauty and bodycare

Increasing the perceived value of products such as toiletries and cosmetics is almost an art form in itself. A knowledge of the materials and processes that can be used to raise this value is essential to both the design and marketing. Italian designers are renowned the world over for their design sense so it is no surprise that the Italian people are spending more than ever on beauty and bodycare products. Approximately two billion of these products were sold in Italy in 1988, through three main channels: 40 per cent through the mass retailing market (department stores, supermarkets and high street drugstores); 33 per cent through specialist perfumeries; and 12.5 per cent through specialist pharmacies.

A small but nonetheless important proportion is accounted for by beauty care institutes, hair salons and door-to-door sales. The interesting point about these ever-increasing sales is that fewer and fewer products are packed in simple glass bottles or cartons: most come in polyethylene or polypropylene tubs or tubes.

One very well known company, Helena Rubinstein, claims always to be looking for new ideas to boost the appearance of its containers. On the top of its Barynia fragrance container, the designers chose to add a large cube of ICI's Diakron, an acrylic plastic, said to outshine glass in clarity and sparkle. The material was selected by Rubinstein in conjunction with French lens-maker Gaggione, who have been working with plastics in the areas of display and point of sale for over 30 years. The material not only looks good, but also has the chemical resistance to withstand any reaction with the solvent in the perfume.

The Ultra range of cosmetics, designed by David Davies Associates, encompasses various packaging materials, including both glass and plastics - but which is which? The high gloss of the plastics containers cannot be distinguished from glass until they are in the hand. Then their lightness becomes apparent.

Medical packaging

Acrylic polymers have also been used to good effect in medical packaging. Diakron for instance is used regularly in the manufacture of blood filters for use in transfusion. The housings for the filters need to be strong enough to withstand knocks and a certain amount of rough treatment and again the chemical resistance is an important factor. The clarity of the plastic is significant, too, for surgeons and doctors can often make important deductions from the colour of different blood samples, or other liquids, as they pass through the filters. Materials that are tainted with impurities, however slight, must therefore be avoided by the designer.

The success of polyethylene terephthalate (PET)

The expansion and use of PET in its various forms has been one of the major packaging success stories of the last decade. Its most important contribution so far has been in the carbonated soft drinks sector. Designers were quick to design other carbonated beverages in PET - and now the take-home trade for beers and ciders has built up considerably, especially in the US and UK, but also in Europe generally. An extra coating of polyvinylidene chloride (PVdC) improves the resistance of bottles to oxygen penetration and gives a shelf life of around nine months.

PET is so successful today that it has become the designer's dream material. Work commissioned by various brewers has shown that textured PET bottles can easily be produced. This reduces the "soft-drinks" image that PET sometimes evokes. One of the first was Metal Box's dimpled, beer-glass look for John Courage's Bitter and Light Ale. Frosted bottles and a ripple effect have also been created for popular spirits.

Mineral waters in PET

Besides alcoholic beverages, mineral waters are also being designed in PET. In the past they have traditionally appeared in glass or PVC bottles, but PET's clarity and sparkle, combined with the fact that it is almost unbreakable, have attracted designers in the mineral water market to look closely at the material.

Mineral water has proved a difficult product to pack in plastics as it has little taste of its own. This means it can be affected by trace impurities of acetaldehyde in PET, which confer a slight lemon taste. However, PET manufacturers including ICI have devised a separate polymer for use with sensitive products such as water, which generates undetectably low levels of the impurity.

The French are the major consumers of bottled water, drinking almost 5,000 million litres annually, closely followed by the Germans (3,500 million litres) and the Italians (2,800 million litres). The UK in comparison is relatively new to mineral waters and annual consumption has yet to reach 75 million litres. But the market is fast growing, creating new design opportunities.

Although glass is a difficult material to beat because it is tasteless, the PET design has nevertheless managed to penetrate the market, especially in Europe, where mineral waters are most successful. In Belgium, for example, Spa is sold in 1.5 litre PET bottles, while the French Evian company is using PET for export markets. In Switzerland, the Henniez brand is sold in one litre PET bottles, with 190ml miniatures for the airline trade.

However, apart from the UK, most mineral waters are supplied in glass or PVC, perhaps the major reason being that in many other European countries PET has to be approved by the local authorities before it can be used for mineral water. In Germany there are also regulations that manufacturers and retailers must introduce workable recycling schemes before any new plastic can be introduced. PET is no exception. Even the mighty Coca Cola GmbH was prevented from launching Coke in PET in Germany until a recycling scheme was in place. However, while recycling the material has been a problem in the past it is not now. The material can now be ground down and separated into its chemical constituents to be recycled in non-food applications, such as strapping tapes, injection-moulded items and sleeping-bag fillers. In the US a large yacht was built from recycled PET Coca Cola bottles mixed with glass reinforcing fibres.

This French pack from Volvic, made from ICI Melinar PET and moulded with diagonal ridges for ease of pouring and to create its own identity, uses the theme of an oasis - in its name and as the illustration - to attract the thirsty buyer. The new packs replace PVC ones and extend the product's shelf life by 3 months.

Packaging wine in PET

Europe is also having to grow accustomed to wine in plastics bottles. Initially only large sizes such as 1.5 litre bottles were available, as the vintners were happy with glass for their premium 75cl sizes. Part of the reluctance may have been the reduced shelf life plastics offer, but PVdC coatings now prolong shelf life for 18 months or more. Now, wine appears in all sizes from 3 litre bulk packs to airline miniatures. A useful characteristic of the plastics bottle is its ability to be moulded in variety of shapes - nesting miniature wine bottles have proven popular with most European airlines. The concave side enables the bottles to stack closely in storage, saving space, and they then lie flat on passengers trays, reducing the opportunities for spillage.

Spirits suppliers are still reluctant to place quality brands in PET, with two exceptions: in the duty-free marketplace and for miniatures. The suppliers are unwilling to say just how much PET saves in cash terms, but they do claim that the weight savings PET offers over glass means that airlines can save enough fuel in a year to fly a Boeing 747 from London to New York.

In design terms, these miniatures offer everything one would expect of a full sized bottle - the labels are suitably high class and the bottles are green. Made from ICI Melinar PET the bottles are light and the vintners are clearly aiming the wines at people with taste on the move.

Bulk containers

Most beverage containers, from 0.25 to 3 litres, can now be designed in PET. Four-litre bottles were produced in the US recently, but showed little signs of being accepted by consumers because they were simply too heavy. In addition, their height made them unstable during the filling process, so they were impractical from the suppliers' point of view also.

The HDPE jerrican, right, for Castrol GTX has two moulded handles - one for carrying and one on the side to help pouring. The integral transparent strip for measuring contents is another useful feature. The familiar Castrol GTX strip in green and red stands out well on the white background of the label and against the body of the canister, too.

However, five-litre containers with special ribbing for added strength are a regular sight now in the agrochemicals market and an additional polypropylene carrying handle helps the consumer carry the unusually heavy bottle. The containers are made in green, amber or transparent PET, although in theory any container colour can be manufactured. One of the first agrochemical products packed in a five-litre PET container in the UK was a Shell Chemicals herbicide, Broadshot, for the control of dock, nettle and thistles. The container was designed with an especially wide neck for ease of filling and pouring and the brown-tinted polymer reduces the deleterious effects of ultraviolet light on the chemicals within. The container is sealed with a foil membrane for safety and also to offer tamper evidence.

High performance

Although the major market for PET is for soft drinks, there has been a continual process of the adoption of PET by designers serving other sectors. Pharmaceuticals suppliers, wines and spirit merchants and ready-meals suppliers have all turned to new, improved forms of PET. Perhaps the most common objective is to improve the performance of PET at high temperatures - standard PET normally melts at the normal temperatures at which foods such as jams or jellies are processed, up to 95 degrees Celsius. Neither can foodstuffs be pasteurized in PET at temperatures of up to 75 degrees Celsius, for instance.

Now, however, manufacturers, such as Sidel in France, have introduced processes that enable PET to be strengthened against temperatures up to about 70 degrees Celsius without losing any of its characteristic sparkle and shine. Processes do exist for treating PET to withstand higher temperatures, but the practical

application of the technology is not yet commercially acceptable. In the US, however, the soft drink Gatorade hit the headlines when it became available in one of the first all-plastics packages to be hot-filled. The 64-ounce container is heat treated to withstand the high temperatures at which this popular sports drink is filled. The bottle with its specially engineered side panels has ribs and a steep shoulder slope that help it withstand tremendous stacking pressure, yet it maintains an attractive modern profile.

Moulded feet

Another development that has made PET bottles more attractive is the introduction of separate feet moulded into the bottle's base, enabling it to stand freely. Previously, only a hemispherical base could be produced by blow-moulding machines and black, injection-moulded base cups of high-density polyethylene were necessary to enable the bottle to stand upright.

PET foods

Foods are the next obvious step for PET and a series of ICI trials in Scandinavia have shown food suppliers and designers how versatile the material is. The inert properties of PET make it suitable for wide-mouthed jars, which can be used for a range of foods, from coffee to sweets. PET jars are now a regular sight in New Zealand, and the UK is also beginning to see prepared foods such as salads and desserts in PET.

Bottles for edible oils are also being designed in PET, where glass and PVC have been the main design alternatives in the past. By the beginning of the 1990s, ICI expects that around half the world's edible oils will be packed in PET, compared with around 10 per cent in 1983.

Away from the bottle format, trays of PET can be taken straight from a freezer at -40 degrees Celsius and placed in a microwave or conventional oven. These trays can also be boiled and usually have a waterproof lidding of clear PET. At present these convenience food trays are produced in plain colours - usually white or buff - but manufacturers believe the technology now exists to produce coloured polymers and to support the printing of graphics on the polyester film lid. This would remove the need for the trays to be overwrapped in high quality carton sleeves.

Some designers may feel, however, that the carton outer provides a much greater opportunity for photography and colour to support the sales message than a simple transparent film. Indeed, although a great deal can be said for enabling the potential buyer to see a product inside a pack, with inherently unattractive products such as frozen or ready meals this can be a positive disadvantage - better to show the promise of the product in its final, processed or assembled form.

Designers have been reluctant to specify plastics materials for foods, partly because of the problems of high temperature filling, but now confectionery regularly appears in HDPE or PET jars. Coffee and jams, too, appear in PET. These wide mouth jars from Lin Pac Plastic Mouldings are in ICI Melinar PET.

Noxious chemicals

Completely outside the food and drinks sectors, PET can also be used for noxious chemicals in Group 1, as defined by the United Nations Committee of Experts on Transport of Dangerous Goods. Before pharmaceutical products can be seen regularly in PET, however, long clinical trials have to be conducted on the various families of drugs and their effects on plastics. However, 100 and 150ml containers are now being produced for the simpler pharmaceuticals, such as paracetamols and aspirins.

Cosmetics and toiletries

In the cosmetics and toiletries markets, European packaging manufacturers and perfume houses are investigating PET as a packaging material. Guy La Roche has already installed PET bottle machinery to pack its Fidji range of perfumes. The material can be coloured and printed as any other plastics material and its inherent strength makes it an almost perfect container. For facial creams and lotions, clear thick-walled containers are needed and a tougher form, crystalline PET, has been developed for this market.

Conclusion

Designers have a wide range of plastics packaging materials available to them today, but it seems that new developments, such as the relatively recent commercial application of PET, can quickly and radically change the face of packaging. A report by Italy's Battelle Institute on advances in packaging technology suggests that the major design advances of recent years, such as high-barrier plastics products, aseptic packaging and tamper-evident closures have come through research, which will continue to transform the packaging world.

PET is almost worth a chapter to itself, to explain the new avenues it opens for the designer. In the US, beverage cans capable of withstanding extremely high pressures (13.6 bar) have been on test for some time, while non-carbonated drinks can also now be packed in plastics cans. In Europe, as always a little behind the US, the market for the microwave oven is in its infancy. The designer of effective, innovative microwavable packaging therefore has many years of opportunity ahead. Coextruded crystalline PET trays are already popular.

The advantages of PET and plastics in general - shape and transparency - are applied in this new perfume called Senso from Ungaro. The whole range of perfume, eau de toilette, soap and deodorant spray are presented in ICI Melinar PET.

These sports clothes from Austrian manufacturers Huber are packed in rigid, transparent PET boxes rather than the traditional cellophane bags. The added quality of the packaging helps the buyer to understand why the manufacturer is charging a premium for the goods.

Sterilization, once restricted to tinplate cans and glass containers - is now effectively possible with coextruded or laminated high-barrier plastics. Other high-tech design applications include controlled atmosphere and vacuum packaging, both plastics specialities.

In short, the future looks exciting for modern-day designers. But to realise the full benefit of today's technology, they must become fully involved in the plastics world, using the material in conjunction with and occasionally in place of the traditional packaging materials. There is a place for all these in the designer's portfolio.

Extending shelf-life, but retaining flavour, has become a major thrust of the modern pack designer. One answer to this rather awkward problem has been the high-barrier plastics bottle. These multi-layer bottles and jars are generally produced by co-extrusion blow-moulding using high-barrier materials such as ethyl vinyl alcohol (EVOH) as one of the layers.

In the UK, Metal Box's Lamicon bottle has been adopted by US food giant HJ Heinz not only for ketchup, but also for salad cream and a reduced calorie salad dressing. The material's typical five layer structure consists of water resistant, inner and outer base layers - usually of polypropylene or polyethylene - and a central high barrier layer that resists the passage of oxygen and essential oils.

The increasing range of barrier materials available to the designer, together with the option of hot-filling, means that these high-barrier containers can be used for many products for which plastics were previously unsuitable, including foodstuffs such as ketchups and mayonnaise. In addition, the versatile blow-moulding process enables custom shapes to be generated to promote the customers unique brand image.

Heinz, for example, has introduced a new style 460g container, to bring the UK version into line with the US product. The bottle is a slim keg shape incorporating prominent 'eyebrows' to enhance the distinctive label shape. In addition, the containers are embossed with two '57's above the eyebrow - reinforcing the company's famous '57 varieties' trademark. The new salad dressing packs use different eyebrow shapes to accommodate the more oval labels for the products.

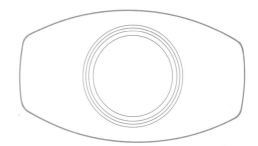

Heinz' new UK design shown opposite - bringing it into line with the US.

Using the versatility of the blow-moulding process, European retailer Asda has moved away from the traditional long-neck look of high-barrier bottles and has adopted modern eye-catching graphics for this range of English dressings and French mayonnaises.

Iceland picked this range off the shelf for its own-label range of ketchups and sauces.

Cawston Vale relaunched its range of concentrated fruit juices in a white 500ml Lamicon bottle designed as an adaptation of a standard bottle known as the Boston Round and having a year's shelf life. The concentrates were previously packed in clear monolayer plastics bottles offering a shelf life of only four months.

Many more companies are now choosing the high-barrier route to product longevity. Lamicon bottles, available in the US from the Continental Can Co, have also been introduced by Hammond's for a 350g American-style ketchup and for a complete line of sauces and mustards. The shatterproof bottle offers a shelf life of 12 months and resists the temperatures of the hot filling process. The surface of the bottle is bright, suggests the company, although, as with many barrier bottles, it seems to dull the colour of the product slightly.

A 750g version of the bottle, with a foil membrane over the mouth, has been developed for the Danish Beauvais ketchup brand, and this has extended its shelf life to 18 months. Benedicta, one of France's leading sauce makers, has launched two new salad dressings in Lamicon bottles: a low-calorie mayonnaise and a salad dressing called Sauce Crudites Nature. The flask-shaped pack incorporates the now traditional long-neck section, which the company plans to use for on-pack promotional sleeves, and a wide main face for a large, often shaped label. An interesting departure is the use of coloured bottles, rather than the more usual translucent container: the mayonnaise is in a pale yellow bottle, while the Sauce Crudités is green.

The designer eager to use this pack for the versatility of the moulding process, or the technical benefits of the high-barrier material may have difficulty initially in persuading clients to take this route, because of the added expense of the material and the production process. But Metal Box, now part of the French CMB organization, believes that although the Lamicon bottle costs more than glass at traditional retail sizes, the reverse is true at catering pack sizes. It adds that market research shows most retailers are able to absorb the price premium because of reduced handling costs - the bottles are significantly lighter than glass - and because of the lack of breakages. Metal Box is quick to point out that the new packs will not be instrumental in the demise of the glass bottle and jar, but clearly competition will be very strong.

ICI has also expressed interest in the high-barrier bottle and is actively working on new materials in conjunction with food manufacturers. However, it believes that in markets such as coffee, toiletries and pharmaceuticals, barrier plastics will have to offer an even better design. This is principally because of the ever-widening application of polyethylene terephalate or PET.

*3.***3**

Glass containers

Until the tin can was invented in the eighteenth century, glass containers – bottles and jars – were the unrivalled solid containers for foods or chemicals and indeed for storage generally. The rudiments of glass-making has been recorded as far back as 7000BC, perhaps arising as an offshoot of pottery, and the first glass-making industry was established in Egypt by about 1500BC.

The reason glass containers could be made by early people was that its base materials - limestone, soda and silica or sand - were in plentiful supply. When these were melted together they fused to form a clear glass that could be moulded while hot.

Glass has long been a staple medium for wine bottles, but designers have endeavoured to use the mouldability of the medium to its fullest extent, especially for the larger volumes. This range by Mittelman Robinson incorporates an integral handle and neck ring. Note also how the label shape parallels the shape of the slightly unusual body.

Characteristics of glass

Glass is superbly strong and even the weakest container can carry a dead weight of over 100kg, although it has a low impact resistance, and shatters easily if dropped. An important factor in the biggest market for packaging - food - is the ability of glass to protect contents from contamination.

Another increasingly important facet of glass's use as a food packaging medium is that it can resist high temperatures and be placed straight in the microwave oven. With the accelerating trend towards convenience foods, this means designers should begin to consider using glass as an alternative to canning.

Working with glass

As with any pack medium, designers should consider every aspect of working with glass, but as glass is such a popular pack, there are already guidelines to follow when drawing up new pack proposals. Bottle or jar capacity is usually expressed as the volume a container should hold when filled to a certain height; this is an average value and in fact the bottles may contain less or more, within certain tolerances. In some cases, there may be a legal requirement to emboss the capacity on the bottle itself.

Considering the type of substance to be packed, the designer should obtain a sample early on to assess the product's appearance in relation to the container. He or she should establish whether it will be packed cold or hot, as glass expands, changing size dramatically. The chemical properties of the contents may affect the choice of closure as some plastics are rotted by strong acids like vinegar.

Temperature and pressure

Bottles and jars must resist thermal and pressure shocks, so the designer needs to appreciate the details of temperature and pressure changes likely to occur during the processing or storage of the pack. These can occur during washing or filling, for example. Carbonated drinks can increase bottle pressure to more than 5 bar during filling, while beers can raise pressure in a bottle to about 6 bar during pasteurization. A vacuity or headspace between contents and cap is usually allowed to account for the expansion of liquids at different temperatures. When any sealed container is heated, its internal pressure will rise, and the extent of the rise can be controlled by providing a large-enough headspace.

Caps and closures

The appearance, colour and decoration of the closure or cap should also be considered an integral part of pack design. Cost may restrict the choice available, so the designer should budget carefully.

For high quality, glass is the medium to choose. In this Vecchia Romagna bottle, designer Michael Peters Group incorporated a spiral neck twist and a solid, gold medallion in a central dimple to replace the label. The glass stopper adds another quality touch.

Production considerations

In addition to design criteria, the designer must be aware of the machinery that will be used to manufacture and fill glass containers. The necks and shoulders of bottles may need to be grasped by palletizing or filling machinery, so should not be too delicate, or too angular for the robot suction cups to grasp. As far as moving bottles and jars on conveyors is concerned, cylindrical containers or bodies with flat sides are the most stable. However, waisted or recessed bodies are satisfactorily stable, provided that there are two different heights on the body at which the diameter or width is equal.

Labels

The size and shape of intended labels should be considered, too. The best labelling surface is cylindrical, with a large radius, so that the label can be smoothed round the curve in a simple wiping action. Spherical and concave surfaces are largely out of the question as paper easily wrinkles when folded in more than one direction. Plastics sleeves could be used here, but are not at present. Sleeves can be shrunk onto a container by heating different parts of the label, causing it to take up the shape of the bottle without creating wrinkles.

Coloured glass

Coloured glass may be chosen for traditional reasons, as with wine, or to protect sensitive products, such as photographic chemicals and solvents, from the light. The designer should check with the customer at the various briefing stages whether colour is an important factor, or whether the designer has a free hand. A new technique for colouring glass was introuced in 1988 by a division of UK's Associated British Foods, under licence from an Australian research company, Vapocure. The colour or colours are applied as a coating, from a spray gun, rather than within the glass. As a bonus, the coating strengthens the container, claim the companies. International Distillers and Vintners has used blue-coated flint glass for its Bombay Sapphire gin, most of which is exported to the US.

The added weight of glass over plastics containers helps consumers feel they are getting something for their money. Almost all olive oils are sold in glass for this reason. Note the tall thin shape of this bottle, an effect which the thin label enhances.

Recycling

Recycling is of major importance to makers of glass containers, as it can reduce their energy bill dramatically. It can also be used, in today's climate of concern about the environment, to give glass a competitive edge over plastics packaging. While no concrete market research has been done on the acceptability to the consumer of the returnable bottle, it is worth emphasizing the recyclability of glass. With some products, then, such as beer, cider, minerals and some dairy produce, the designer may wish to use returnable bottles. These must be capable of withstanding repeated rough use with the minimum damage. The economic justification is that the cost per trip - between store and home - should be small. In the UK, the doorstep delivery and collection of glass milk bottles means that the containers can be used up to 30 times, bringing cost per trip to around 1 pence. The alternative is reducing the weight and the strength of the bottle to the minimum needed for one trip. Non-returnable containers should therefore be about two-thirds the strength of a returnable bottle.

Handling characteristics

There may be special holding or pouring requirements the designer must incorporate: will the product be opened with wet hands, in a bathroom, kitchen or garage? How much of the product will be used at once? How does the product pour? Thick products such as syrups or oils need a wider mouth.

Legal requirements

Finally, there are always legal restrictions or requirements with which glass designers must comply, particularly those concerned with weights and measures and the use of bottles as measuring containers. Apart from a general liability under local sale of goods acts, and health and safety at work requirements, there are several specific requirements to be aware of: control of bottle contents, the packaging of poisons, and packaging for export are just a few. Other points include avoiding patented or registered designs. Designers should build up a library of material to help them in this.

The glossy black and red label on this bottle, together with the neck card, enhances the high quality of this thick-walled Italian spirits bottle redesigned by Mittelman/Robinson. The ridged sides and embossed shoulder logo add to this impression.

Designing with glass

Plastics bottles, particularly PET and PVC, do not have the surface qualities that many designers regard as important for their products. If a designer makes a glass jar or bottle frosted, it can have the appearance of being cool and refreshing. Some manufacturers have been able to give PET this finish (see Chapter 3.2), but it is generally not common in plastics. Another glass finish is marbling, used effectively by West German glass designer August Heinz for cosmetic containers.

Other design advantages of glass is the wide variety of effects that can be achieved in glass, either through shape or surface finish, such as the impression that a jar is tightly packed with product. Jams and preserve makers achieve this through using bulging sides, or barrel-shaped containers. By providing a handle, or a special pouring or delivery spout, a product can be made to look useful or "handy". Salad dressing bottles shaped to fit the hand also join this category, as do the pouring carafe and some medicine bottles. Flat faces or facets put forward the high quality image, reminding consumers of jewels or crystal glass. This effect is useful for cosmetics and expensive spirits.

This bottle's shape - reminiscent of a cricket bat - helps support the cricket image and offers flat faces for the screen-printed label (see page 65).

Glass offers the designer an effective way of providing a range of products. The liqueur bottles, below, range from 2 litre down to miniature size, but the shape of the cap and the colouring and style of the labels ensures they are seen as the same product. The old design on the left, using a solid metallic pink was perhaps too sickly for the modern taste.
(Mittelman/Robinson)

Shape

As discussed in Chapter 1.4, shape itself can be used imbue characteristics to a package: old-fashioned shapes can suggest reliability, maturity and the fact that a product is well-established. By using streamlined and angular designs, in comparison, the pack can be made to appear modern. These ideas do not apply to glass alone, but when used in conjunction with this transparent, heavyweight material, they can become very forceful marketing tools.

Size and capacity

At present manufacturers have a fairly free choice in the quantities in which they can sell their products, which means that designers have an equally free choice in the sizes of container to use. However, national legislation, such as the UK 1963 Weights and Measures Act prescribes mandatory size ranges for milk, instant coffee, honey and jams, and it seems likely that more products will join the list. In the UK, complete bottle specifications are produced by the British Standards Institute, among others, to cover all dimensional and physical characteristics, with tolerances and means of verifying them.

The list of general trade bottles available to the designer is extensive, but there are a number of main groups including winchesters and oval section bottles, usually fluted for poisons. Spirits offer another example, generally comprising tall round bottles for 750ml or more, with flasks for some of the smaller sizes and various designs of miniature. Of course, there are notable exceptions: Johnny Walker Scotch Whisky and Jack Daniels Bourbon use square bottles, which are more expensive.

Meeting the challenge of PET

Over the past decade, major developments have been made in the plastics markets as new products such as oriented polypropylene (oPP) and new mixtures replace tried and trusted traditional materials. In worldwide confectionery and biscuits markets, for instance, bright, airtight, oPP films have almost completely replaced paper and cellulose films.

But it is the development of polyethylene terephthalate (PET) bottles, using material from ICI and Eastman Chemicals among others, that has all but wiped glass from many of the shelves at the supermarket, as described in the previous chapter.

Plastics are not having market trends all their own way, however, for designers are rediscovering glass. A crucial year for the comeback was 1985, when an international consortium to research glass strength was established, the International Partners in Glass Research. The group incorporates US and European companies with the same aim: to see glass ten times stronger, yet only half the weight, by the mid 1990s.

Other members of the consortium include ACI International of Australia, the US's Emhart Machinery Group and Brockway (now merged with Owens-Illinois, USA - parent of United Glass, UK). Another North American partner is the Consumers Glass Co, Canada; West German industry is represented by Weigand Glass; and Japanese industry by Yamamura Glass.

Stronger and lighter

The plan of the International Partners in Glass Research consortium is to build on the knowledge already gained in other areas of glass technology to develop techniques of preserving the inherent strength of glass: when glass first drops into bottle moulds, as gobs of molten liquid at 1,400 degrees Celsius, computer calculations show that it is twenty times stronger than stainless steel. However, it soon loses this strength as it acquires microscopic flaws at the surface.

Finishes

The knowledgeable designer, in conjunction with the glassmaker, can now specify the surface finish of containers. Most bottles and jars, for instance, can now be spray-coated with a variety of compounds containing titanium or tin. The different coatings harden the surface of the glass to different degrees, but both help to prevent the bottles being scratched excessively. These coatings are not that new, but scientists at the Massachussetts Institute of Technology believe that new chemicals called sol/gels can also be used as coatings. Sol/gel coatings are special because they do not simply paper over the cracks - they may actually be able to reseal flaws and scratches. This is only one area of research among thirty or so separate projects.

Glazing

Today's biggest packaging design opportunity for glass has actually been practised for thousands of years: glazing. Like pottery, glass containers can also be glazed. A cocktail of special chemicals is bound firmly to the surface of the bottles by the heat of an oven. The process is very similar to the way pottery is glazed in a kiln. The glaze compresses and hardens the bottle. These developments are being researched around the world and in local trials over the past few years some of the bottles on our supermarket shelves have become two to three times stronger than they once were. The manufacturers are not yet prepared to reveal how those tests fared, but clearly the results will interest every designer interested in glass as a pack medium.

Weight

As a by-product of the search for strength, it is likely that the glass bottles and jars of tomorrow will be half the weight they are today, which will make bulk buying in the weekly or monthly shopping trip much easier for the consumer.

Owens-Brockway, a subsidiary of United Glass, has already announced weight savings: it has designed containers that weigh up to 20 per cent less than previous containers. The results of the new design were first seen as early as 1986, when the company introduced its 75cl wine bottle, which weighed only a fraction more than existing 70cl equivalents. The bottles were introduced for new European legislation aimed at rationalizing the 75 and 70cl sizes by 1989. The company had also experimented with special coatings to strengthen thin glass, but the major contributor to the lighter weight of the wine bottle was greater control of the manufacturing process. By introducing computer control at its glass furnaces, United Glass has provided greater flexibility and more precision in the preparation of molten glass for the furnaces. This directly affects the quality of the final products.

However, these considerations need to be balanced against the use of the weight of glass as a marketing advantage, where it can give a better quality "feel" to a product.

Glass moves upmarket

It would be true to say that designers' use of plastics for soft drinks and squashes has forced other designers to shift glass upmarket as a pack - the results can be seen in the way that liqueur and spirits distillers are devising ever more ornate styles to match the high-quality image they advertise for their products. New forms of pack, such as the wine carafe, have enjoyed increasing success in glass, too.

One of the more exciting glass packs of recent years was a range of pineapple-shaped jars from Dole in the USA, for pine-apple sauce, reputed to be the first time pineapple had been packed in glass. The jar's manu-facturer and designer, Owens-Brockway, took advantage of the mouldability of glass to come up with raised helical florets, ranged round the outside of the 16, 25 and 35oz jars. The idea of the presentable jar makes the pack not simply a commodity, but something a consumer would put straight from cupboard to table. The pack thus gains a high profile and becomes synonymous with the food. And once the food is finished, the container may also be used as storage in the home, as a permanent sales message to the consumer. In 1988 Rockware Glass produced 75cl green bottles for Hedges & Butler's French version of Eisberg, a low alcohol wine. The key feature of the design was a stippled effect achieved by embossing a series of mountain peaks about two-thirds of the way up the bottle.

This 1.5 litre bottle from Grants of St James's incorporates an integral handle for ease of pouring and a resealable screw-cap closure in tinplate. The label uses subtle typo-graphy to keep standards high, even though the container uses bulk product tactics. (Designed by Brand New)

Versatility of glass

The ability of glass to be moulded makes it very versatile: it enables jugs with integral handles to be made for large, yet small-necked bottles. This mouldability means that glass can also be fashioned into quite delicate items, such as pharmaceutical ampoules and other forms of drug container. These fragile items usually require some form of secondary package, such as the Dividella divided cartons for drug ampoules.

Another benefit of glass packag-ing that consumers seem to appreciate is that it is re-usable and can perform a range of func-tions. Of course, jars and bottles have been re-used for many centuries, but many manufactur-ers are now producing tumblers with plastic lids for mustards, jellies and preserves; one German yoghurt manufacturer offers cork stoppers to turn glass containers into storage jars for herbs. The key message in these cases is that glass is good value.

In fact, designers should use the fact that glass projects this "wholesome" image: it does not corrode; it does not stain or leak; it remains attractive in use; it is imper-meable. Glass should be used for high-quality goods. Furthermore, it can easily be filled with products at very high temperatures, because glass itself only becomes liquid at hundreds of degrees Celsius. The problems of early containers that cracked with the temperature shock no longer affect most high-quality glass containers.

Glass jars can also be stacked on the retailer's shelf without being crushed. They are easily re-sealable. There are many good, practical reasons why products should be packed in glass, but also important are the perceived characteristics such as purity and visual appeal: the product can be seen inside and the consumer knows or believes that it is unaffected by the packaging material, whether the product is a food, a cosmetic or a drug.

Applications

Despite the onward march of materials such as PET, and PVC for squashes and cordials, glass is still an important pack. This is especially so in the USA, West Germany and France, where the use of glass containers for beers, cider and wine is more wide-spread than in the UK. Nowhere is this more plain than in the massive French hypermarkets, where almost all beer appears in 25cl non-returnable or one-trip green bottles. Even soft drinks containers up to 1 litre appear in glass in France, despite heavy increases in PET usage. There are several smaller companies such as Pochet, which appear to specialize in supplying small glass bottles to the well known French perfumery and cosmetics houses. One of the leaders in this field as well as in pharmaceuticals, is the Saint-Gobain subsidiary Desjonqueres.

Glass's main uses in the UK appear to be freeze-dried or instant coffee, and jams. Within the food area, spreads and pastes show little sign of moving into plastics, but they only account for a small market. Pharmaceuticals are rapidly appearing in plastics and cartons because they are lightweight, but it appears that it will be hard for glass to be ousted in the cosmetics market.

These jars from jam maker Hartley have been almost the same slightly tapered shape for years, but in the face of heavy competition from many own-label brands the designers have been forced to use the label to shake off the competition by moving the product upmarket. An oval, rather than rectangular, label, showing 4-colour reproductions of the fruits the jams contain lifts the pack out of the ordinary. (Michael Peters Group)

Cosmetics

History has it that the first glass blowers were the Syrians, who among other things made scent bottles in stylish blues and purples. The Romans embellished the art and devised new ways of cutting and engraving glass, also founding early mass production techniques throughout Europe. Even then, perfumers realized that glass lends a quality in value unmatched by other materials.

According to UK pack designer Packaging Innovation, only in glass can the different effects be achieved that differentiate one fragrance from another in a highly fragmented worldwide market: "Some products exploit the crystal clarity of transparent glass in strong geometric shapes to create a faceted, jewel-like appearance, such as Gem and Giorgio, both relying on light transmission for their effect." Calvin Klein's Obsession is designed in a comparable fashion, using curves that give the impression of a pool of light in the low smooth shape of the pack.

Acid-etching

In the late 1980s, the process of acid-etching became popular. Three new products used the technique to great effect: Montana, Salvador Dali and a range from Jean Patou. Taking only one of these, Montana is a bottle designed in spiral, translucent steps. The last step is in fact the bottle closure, with the name in metallic blue etched into it. The design is memorable and unmistakable.

Many perfume bottles use heavy glass stoppers. The effect they achieve is unmatched by simple plastics screw caps, whether or not designers use new metallizing techniques to improve their appearance (see Chapter 3.2). Shape and outward design is essential in this market - no two perfumes can afford to look the same. But there are many imitators of good design, such as the traditional sleek lines of Chanel No 5, or Yves St Laurent's Opium. Opium is interesting from another angle: it uses a plastics window in the product which echoes the feel of a jewel trapped inside.

The world's leading perfumers are the French and the Italians, and the quality and feel of their designs are unsurpassable, as can be seen in any of the collections from Luigi Bormioli. Japanese designers, too, are important cosmetic players although they are using PET to good effect, offering a new, light feel combined with glass's traditional high gloss and clarity. Tinted PET can be produced in very subtle colours.

Creams, powders and pastes tend to be packed in jars, which traditionally are glass or ceramic. But here too plastics are being used with great effect to approximate the look and feel of the heavier materials. For expensive products, though, nothing will replace glass for many years to come. The added weight factor is a major design feature here.

Most glass manufacturers have a range of stock bottles from which designers can choose, rather than establish their own shape - which is more expensive. These stock bottles, labelled and decorated with ribbons offered Penhaligon's perfumery exactly the balance it wanted between modernity and tradition. The glass stopper rounds off the image. (Michael Peters Group)

Wine

Most wine bottles, too, are of a traditional design. The shapes and dimensions for hock, burgundy and claret bottles are already effectively standardized. Some wine bottles use coloured glass to distinguish between regions. Riesling wines all tend to appear in tall, slim bottles, but a green bottle tells you the wine came from the Mosel region, while an amber or brown bottle is from any region bordering the Rhine. Blue bottles once indicated wines from the lesser known Nahe River region, but their use is now unrestricted.

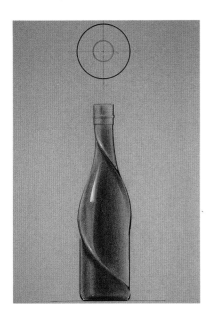

The versatility of glass is almost unrivalled in the packaging world - which is why designers today go to extraordinary lengths to make their products look different from the next one on the shelf. Spirals and ridges have all been appearing on the supermaket shelves of late. Besides the innovative shape, this Savonne bottle, devised by Gerstmann & Meyers, offers new opportunities for decoration. The diamond shaped label between the two ridges attracts the attention because it is unusual.

Conclusion

Quality and value in virtually all product sectors are now almost always taken for granted. They are no longer the key bases for retail competition. Consumers are concerned with health and health products, for example, and the number of connoisseurs among consumers is increasing. This means that the future of high-quality foods and products, in high-quality packaging such as glass, is assured.

At Verona's annual VinItaly show, the glass bottle still appears dominant in Italian wine markets, although cartons are slowly making inroads for low-priced brands. However, Italian exporters say they cannot sell cartoned wine abroad and as cartons only account for 2 per cent of the Italian wine market, there seems a great resistance from consumers also.

There is some consolation for designers working with carton manufacturers: the tinplate can fares even worse in Italy and plastics bottles are almost absent from the market, except for bulk containers over 5 litres. This is because the Italian vintners feel they would need a huge investment to initiate plastics packaging on an effective scale. As far as bag-in-box packaging is concerned, Italian designers are certainly interested, although the question of flavour retention will have to be addressed more clearly before it gains any great favour. At present the Italians see bag-in-box only as a method of packing wine in bulk.

The great feeling at the show is that wine has always been in glass and should stay that way, which is something designers need to take into consideration.

CASE STUDY PASTA POUR-OVERS

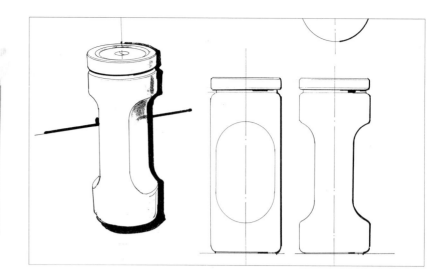

Thanks to co-operation between a glass manufacturer and two designers, Chef Tell's packaging for its Pasta Pour-Over sauces has proved a success. The four products in the range have visual appeal, due to seven-colour printing on a three-quarter length, PVC shrink sleeve, but the appeal is reinforced by the pack's ease of handling. The Wheaton Glass bottle was designed from the outset for the microwave. Once the closure is removed the flat-sided jar, with upward tapering sides that are easy to hold, can be placed directly in the oven. The sauces retail for a relatively expensive $2.49-3.49 per 12-ounce jar, so the pack had to reflect the quality image.

The product is marketed by Madison's of Larchmont New York, which said: 'The glass jar is part of the packaging approach. But the overall look is what we were striving for. We wanted something unique, substantially different from a stock bottle.' Madison's designed the bottle in conjunction with New York agency, Apple Design Source.

The pack is a triumph of design over traditional conservatism, suggest the designers. 'Most of the glass companies that serve the food industries were not encouraging when we talked about the shape we wanted,' said Madi Ferencz of Madison's. Barry Seelig of Apple Design Source agreed. 'The large ones didn't want to get involved in a short-run, custom container. They tried hard to talk us into stock containers.'

Each of the four sauces in the line is a distinctive combination of ingredients and spices designed to be poured over freshly-prepared pasta. 'Microwaveability was a big factor in determining the shape of the jar,' said Seelig. 'A standard round jar can be difficult to grip with a potholder or oven mitt. Our jar has flat sides that taper upward to a flange so it can't easily slip out of one's grasp. That's why it's oval at the bottom and then flares out to a round top.' By eliminating the round shoulders common to other sauce jars destined for the microwave, the chance of bubbling or splattering in the microwave are reduced, adds Seelig.

Because of Wheaton's experience in cosmetics and unusual shapes, it was very receptive to working on the sauce jar and understood what Madison's was trying to achieve. A spin-off for Wheaton Glass is that it hopes to become better known in the food industry and attract further speciality container business.

Apple Design Source supplied the engineers at the glass manufacturer with sketches and a lucite model, from which the Wheaton Glass engineers produced the finished specification.

The PVC shrink sleeve covers the bottle from top to bottom, with a central vignette of Chef Tell, a TV personality, set against a transparent background, through which the sauce's colour and texture can be seen. The other main background colour is black. 'Colour played a critical role in conveying an upmarket or upscale look and feel,' explains Seelig. 'We were convinced that a black background, once taboo in food packaging, gave the packaging a touch of elegance, would accentuate its high-quality image and offer a strong shelf presence.'

The label also acts as a tamper evident seal - just below the skirt of the cap the label is perforated. When the cap is twisted the top part of the shrink label breaks away with the cap. Finally, said Seelig, the visual presentation of the pack becomes a major communicator for the product itself.

The finished result: only a PVC shrink sleeve could have accommodated the shape of the jar.

*3.*4

Metal containers

An army, it is said, marches on its stomach. No one can have known this better than French General Napoleon Bonaparte, who in 1809 offered the fantastic sum of Fr12,000 to the person who could preserve food for his army. The reward was claimed by Parisian chef and confectioner Nicholas Appert, who showed that food packed in sealed tin containers and sterilized by boiling could be preserved for long periods. In Britain, Peter Durand obtained a patent for the tinplate can about a year later and to this day, sterile cans of tinned or bully beef and carrots have been a staple diet of soldiers everywhere.

Tradition dictates that boot polish comes in shallow, circular tinplate containers, although they have become decorated in full colour as printing techniques have improved.

In the past, enamelling or inlaying with another metal was the most popular method of decorating containers that would "appear in public"

Historical records also tell of early cans made from iron and coated with tin in Bavaria in the fourteenth century and how the secrets of manufacture passed slowly across Europe until they reached France and the UK early in the nineteenth century. The techniques then passed to the US courtesy of William Underwood, who emigrated from England to Boston a decade later. Soon, however, the exchange of steel for iron and better manufacturing practices around the world led to a substantial increase in tin can output and a great improvement in quality.

Replacement by plastics

Today, however, things have changed. The army, for instance, ever keen to reduce the weight of the standard issue backpack, has decided that the tin can has had to go. Now most Western infantry carry special boil-in-the-bag foods, such as beef stew and dumplings, curries and pasta meals. The menu offered to the late-twentieth-century soldier is much wider than ever before, is - allegedly - more nutritious, and can be cooked more quickly.

These are just some of the reasons that the ordinary consumer has also been faced with fewer canned foods in the supermarket. However, recognizing the problems facing cans in the retail war, designers are devising ever-more exciting cans, for various applications. US astronauts, for instance, took special cans of Pepsi Cola and Coca Cola on board early Space Shuttle flights. The space-age

cans were designed to cope with zero gravity and took almost a year to develop. Before the launch a NASA spokesman said: "The astronauts will be looking at carbonated drinks dispensing in space, but they will not be running a taste test!" The work that went into putting a can in space will be adapted to improve the shelf life of terrestrial cans, according to US can makers.

Family packs of cream crackers or water biscuits have always been packed in roughly square, airtight tinplate containers. The modern counterpart of this pack may be over-wrapped in a plastics film, rather than bear printing directly.

Metal containers with a good seal keep out moisture as well as air, hence their use for confectionery before the cartonboard box or plastics bag were invented.

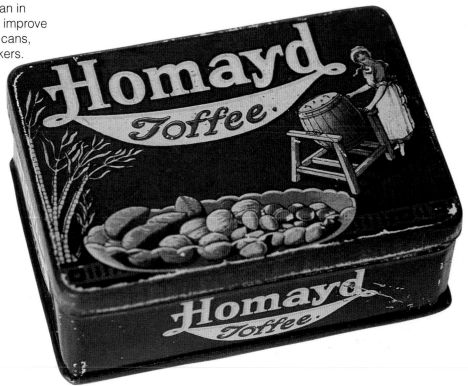

...national ...-operation

In the UK, British Steel has invested £200 million in its tinplate packaging operations and is co-operating actively with high-technology overseas manufacturers such as Hoogovens of Holland and Rasselstein of West Germany. Already the steel soft-drinks can has been reduced in weight by well over a third - from 57 to 35g - while the weight of the generally lighter aluminum can has been cut by over a fifth, to about 19g. In short, the packaging designer has to know why and how the modern can sometimes offers the best pack, and for which products. The key seems to be that technical developments in can manufacturing can overcome the perceived market criticisms of tinplate and aluminum.

Of course, cans these days can be made from paperboard and plastics, too, but it is the metal container made from tinplate and other metals, chiefly aluminum, that will be considered in this chapter.

The "tin" can

The tin can is today a misnomer, for cans are made not from pure tin, but from tinplate. Even tinplate has changed over the years, but modern tinplate is formed from a thin sheet of steel covered with a very thin coating of commercially-pure tin. The steel gives the can strength, while the shiny and attractive appearance, as well as resistance to corrosion, are due to the unique properties of tin. A material increasingly in use in packaging is tin-free steel (TFS), devised to make the "tin" can a more cost-effective package.

For distribution and storage of food, the can is one of the best packages ever designed. Most other forms of food container need refrigerating to keep them fresh and electricity charges are usually the largest expense borne by supermarkets - second only to wage costs.

Changing applications

The use of tinplate has also changed with time. The bulk of it is now made into hermetically-sealed, sterile containers for food and beverages, as opposed to household boxes, plates and pots. A considerable quantity is also used in aerosol and aerosol-type containers for air fresheners and toiletries, and these will be covered later in the chapter.

A shrinking proportion of tinplate - perhaps 5 per cent - is used to pack paints and polishes. Tinplate also plays a significant role in the manufacture of caps and closures for wide-mouthed glass jars, although this, too, is a shrinking market as plastics screw-cap closures have been introduced widely.

Quality foods

There is, however, a problem with tins from the marketing viewpoint. Tinned products are intrinsically dull, with a poor quality image that does not really fit into the pattern of adventurous eating the Western world has embarked on in recent times. An encouraging sign in recent years has been the increased sales of well designed, upmarket, quality canned foods, including gourmet sauces and soups. But market research conducted by international advertising agency Ogilvy and Mather showed among other things that many people feel guilty about using canned foods: these are not seen as fresh or nutritious. Some respondents felt that using canned foods was lazy. So the can could be fighting a losing battle unless designers dispel these fears and persuade customers with positive label designs depicting healthy, modern values. There is a major marketing job to be done and packaging design has a key role to play.

Developments in tinned foods

- *canned vegetables*

  flavour and texture improvements

  now offered in brine or water

- *canned fruits*

  offered in syrup or natural juices

  increasing varieties

- *desserts*

  lower calorie variants

- *meat products*

  improved quality and variety

- *almost all products*

  elimination of additives: particularly salt, sugar and 'E numbers'

Down-gauging

Each tonne of aluminum used for canmaking is now yielding more products than ever before, thanks to designers new ideas and research into the production of thinner and stronger containers. This trend, known as down-gauging, has seen the wall thickness in new can designs reduced by 2-3 per cent each year since the mid-1980s. Canmakers now produce 7,000 more can bodies from every tonne of metal than they did ten years ago.

Designers must follow up these latest trends to provide customers and clients with the most cost-effective use of materials, especially for more traditional metal containers. Household foil is now rolled from a special aluminum alloy which has meant a 15 per cent reduction in foil thickness, over the past decade. This in its turn has meant a reduction of about 20 per cent in the thickness of semi-rigid foil containers as used in the take-home foods and TV dinners markets.

Tinplate containers have a traditional feel about them and also suggest quality. This canister of biscuits, designed by David Davies Associates, in black and royal blue is sufficiently unusual to give the product a head start.

These two-piece aluminum cans of cat food designed by US designer Lister Butler offer a practical wide opening for ease of serving. There is a ring pull so a can opener is not required.

Can making methods

From the designer's perspective, the main difference between cans these days is whether they are made from two or three pieces of metal. Originally the open-top can was made from three pieces of tinplate: two circular ends, crimped on in one of various ways, and a rectangle, formed and welded into a cylinder.

Then the two-piece can was devised in the 1970s, through a process called drawing and wall-ironing. The technique takes a thick circular blank and in a series of three steps draws the wall upwards to the right height and thickness - between 0.1 and 0.14mm. The rim is then trimmed and it is at this point that the can may be printed, but if the can is to be decorated by labelling that will probably occur later, after filling. The interior is then lined, if necessary.

Manufacturing techniques are constantly being refined, as two-piece cans have several advantages over the soldered three-piece can. First, there is a more effective join, or double-seam, at the top-end of the body,

between lid and can. This is because the cylindrical can body does not contain the sensitive seam junction of the three-piece can. This is particularly important for processed foods and carbonated drinks packed under pressure, where the seam is a potential weakness. Second, the two-piece can uses significantly less metal and is therefore cheaper. It appears also, in market research tests, to be more attractive to the consumer, and it has ecological advantages in that it aids the recycling process - no solder appears in the re-smelted mixture.

A second technique of interest produces the drawn and redrawn can, where a wide-mouthed can is first produced and this is then redrawn with a narrower mouth, making it taller. The advantage of this process (DRD) over the draw and wall-ironing (DWI) process is that pre-lacquered tinplate can be used, which cuts costs. The cans made by the DRD process are generally shorter than those made by the DWI process. There are other production economies, too, which the designer should follow up with local manufacturers.

The trend towards lightness

On the whole, the future trend is towards cheaper, lighter weight cans, by the reduction in thickness of the can walls. Soon only a drink's carbonation will prevent it being crushed on the supermarket shelf.

The thin-walled metal can poses no problem for carbonated drinks, since the pressurized carbon dioxide stabilizes the container, contributing significantly to its strength. To test this strength, try squeezing a beverage can before it is opened and then again when the pressure is released.

However, when a can contains flat drinks such as juices, the container can be damaged very easily. This has prevented the widescale use of cans for juices and diluted drinks. Using a patented cryogenic method developed by the German specialist Messer Griesheim, inert liquid nitrogen is injected into filled cans and vapourized to develop a pressure of about 2 bar. The process does not affect foods and can be used in the canning of hot liquids.

Easy - open ends

By the 1950s and early 1960s, beer cans were increasingly seen, but it was not until the development of the ring-pull aluminum end in 1963 that the true potential of the beverage can came to be realized. The effect of the "easy-open" end on the sales of soft drinks in the late 1960s and 1970s was tremendous.

There are two kinds of easy-open end: the pouring aperture, generally pear-shaped for ease of pouring liquid products - oils, fruit juices and beers; and the full aperture for more solid products such as meats, sauces and nuts.

Although other kinds of opening device have appeared on metal containers over the years, consumers have always complained bitterly about them.

The key-opening scored-strip, for example, round solid meats, such as cured hams and corned beef, and round shallow fish tins, is perhaps the best example. However, many of these tins are produced in areas such as South America where packaging technology is old and the management reluctant or unable to invest in new equipment, so this situation seems unlikely to change in the near future.

The strong branding of the Coca Cola trademarks and graphics brings together the different forms and styles of tinplate and aluminum cans into one cohesive family. For the customer, the two- and three-piece cans and various shapes and volumes disappear behind the brand identity.

Aluminum ends

New aluminum openings are designed regularly. "Presto" is a pouring easy-open end devised by the Australian Iron and Steel Company, but it does not incorporate a separate lever. It has two circular tabs, one smaller than the other, formed into the end panel by scoring almost completely around the aperture circumference, leaving just enough to act as a hinge. From the manufacturer's point of view this end is cheaper to produce than the standard ring-pull and the much cheaper tinplate may also be used.

Screws and nuts as well as other DIY products have recently been packed in shallow tins with easy-open ends, again in aluminum. Tinplate lids have been produced but only of the sort that removes completely from the container.

Tinplate ends

Many attempts have been made over the years to design a tinplate easy-open end with a pouring aperture for oils, say, but so far without much success. British Steel, however, recently launched a new generation of steel easy-open ends with Hoogovens of the Netherlands and Otto Wolf of West Germany. The can was awarded first prize in the Netherlands Institute of Packaging annual award ceremony. If mass production is approved, these could appear on many cans by the end of the century.

Types of can

a   **multiple-friction**
b   **three-piece**
c   **spice**
d   **hinged lid**
e   **oblong key opening**
f   **two-piece**
g   **square-breasted**
h   **aerosol**
i   **oval**
j   **oblong F-style**
k   **meat**
l   **beer, beverage**
m   **flat round**
n   **sardine**

Soft drinks in cans

Despite heavy advertising all year round, a good summer can make or break a brewer's year. Good weather in July, August and September can increase annual sales of soft drinks by 15 per cent or more. The soft drinks sector accounts for perhaps 40 per cent of the sales of canned beverages worldwide, but special mention has to be made of the colas, whose sales in cans accounts for over half of all off-licence volume sales in Europe and North America.

The can's exceptional performance in soft drinks looks set to continue for some time, despite pressure from the 250ml and 500ml PET bottles and the smaller juice cartons. This seems to be for a mix of reasons, including the important marketing punch offered by the surface area of the can. As well, there is the fact that cans from the store refrigerator or chill cabinet are so much more refreshing than a cartoned drink, yet they are lighter than the single-serve glass bottle, which appears only to be holding ground as a container for mixers.

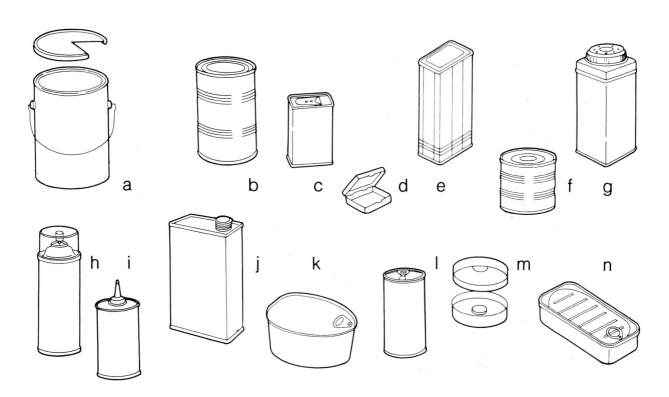

Multipackaging

The multipackaging of soft and alcoholic drinks through the Hi-Cone type of plastics neck, which holds 4 or 6 cans together, is another successful introduction into the UK from the US and is helping to boost sales through all retail outlets. This method of marketing more product is one that designers can use instead of adopting a two or three litre PET bottle. Many brewers use both methods.

Packaging emphasizes premium

In the alcohol markets, lager and wines play an increasing part in the UK - they are already well-established in continental Europe and the USA - and packaging has an important role in emphasizing the high value and premium price buyers can expect to pay. Foil labelling is common in the premium bottled beers market, following German traditions in bottling, and it is no surprise that metallic golds and silver lacquers are also common in can printing.

In the traditional food markets - vegetables, meats, soups and fruits - consumer perception of the lack of freshness and low nutritional value of canned goods has hampered sales. In the meantime, the soups market now faces another opponent: the aseptic carton. Since 1985, Crosse & Blackwell's Four Seasons range of soups has been on the shelf around the world, in Bowater's Combibloc carton, steadily picking up sales. Scandinavian carton maker Tetra Pak has also joined the war, although its cartoning system cannot yet handle the $25mm^2$ particulates that the Combibloc system does.

These two-piece beverage cans, designed by US agency Image for Genesee beers, show the latest techniques in can-making. They are necked-in, to save material on the ring-pull end, and are printed in attractive, high-quality metallic colours. The use of fine stemmed typography enhances the quality image.

Trends in food packaging

Food fashions, such as the re-emergence of fibre as an important constituent of the diet, have re-established baked beans as an important commodity. The health trend, too, together with manufacturers' efforts to produce salt and sugar-free foods, has attracted a great deal of attention among consumers. Designers should not fight these trends, but use them, highlighting the significant contents (or absence) of foodstuffs in on-pack promotional flashes.

Self-heating and self-cooling cans

Two of the most exciting developments in can design of recent years are the self-heating and self-cooling cans. Through chemical reactions in a hollow base these special containers allow hot meals or cold drinks, such as sake, to be presented to the consumer in a safe, convenient pack. One of the first arrived in 1985 when Metal Box announced the launch of Hotcan for the outdoor leisure market. The Hotcan is a 425 gram, single-serve, can within a can that contains calcium carbonate granules and water bags. When the bags are pierced by the opening spike the water reacts with the granules to heat the contents, sandwiched between the two cans.

The manufacturers claimed a shelf life of 18 months for the various stews and casseroles in the range and announced that the cans had been tested by competitors in several ocean races.

In the same way that chemical reactions can produce heat (exothermic reactions), some can also take heat away (endothermic reactions), and it is this that Japanese drink makers have harnessed to pack sake. Hiya Can is an almost standard 360ml aluminum can that contains only 144ml of sake. Inside the can are two smaller containers keeping the chilling chemicals - water and ammonium nitrate - separate. The water container has a thin foil cap, which is broken when the thirsty consumer pushes the bottom of the can upwards. The temperature falls about 15 degrees Celsius to a minimum of about -5 degrees Celsius, depending on the initial, ambient temperature.

With gimmicks such as these self-heating and self-cooling cans, it is important that makers ensure consumers are safe and that the instructions - if complex - are simply explained. It is therefore the designer's task, in the end, to ensure that pack innovations appeal to buyers by combining the roles of product manager, marketeer and designer.

The ring-pull opener is now standard in metal packaging as shown by these cans from the UK and Spain. The symbol denoting the opening method is also instantly recognizable, overcoming language differences.

Quality gift containers

There has been a resurgence of interest in tinplate containers as high-value gift containers and increasingly in the nostalgia markets. The production rate of these specially shaped or coloured containers is obviously much slower than for the high-volume cans aimed at commodity products. For this reason, they usually cost at least twice as much to produce and are recommended only for premium products that can absorb the additional expense.

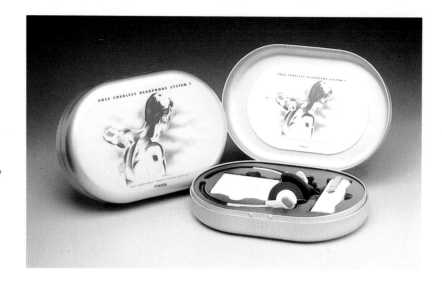

To maximise the impact of the personal stereo on buyers with money to spend, headphones maker Ross Electronics introduced its cordless set in a high-class pearly tinplate container, printed with a modern illustration depicting a young man using the product. The lid inner reproduces the illustration and the headphones are held in place by a thermoformed black plastics tray. (Designed by Michael Peters Group)

Collapsible tubes are today most often made from plastics films laminated to foils or thin metals such as aluminum. But in some instances, such as these alkyds (top right), but also for ointments and adhesives, straight tinplate or aluminum is still the best medium for reasons of compatibility. Note that the tubes themselves have been decorated with the illustrations that also appear on the cartons and bottles of Winsor & Newton inks (see Chapter 3.1).

Using traditional scenes of European cities printed on tinplate coffee containers (below right) conveys the image of exotic blends of coffee far better than a simple textual explanation. The almost nostalgic feel for a Europe of a century ago is weakened though by the use of the resealable plastics lid to keep the coffee fresh, even though the extension of shelf life is an important aspect of design. (Designed by Mittelman/Robinson)

Decorating cans

The decoration of cans offers many marketing advantages. Besides a general improvement in appearance, the direct printing or labelling of cans can have a great impact on product marketing. But it also has an important effect on protecting the packs from corrosion or rust. The specification of printing inks is also an increasingly important aspect. Non-toxic, lead-free pigments are much more readily available today than ever before. Some containers are lacquered and preprinted before they are formed into shape, but a great deal still need to be printed when fully-formed, or "in the round".

Although many designers attempt to keep individual colours separate when printing, to avoid the potential mixing of colours, this is difficult and seriously restricts the creativity of the design process. Some steps towards the easy mixing of colours have now been taken and designers can now be more confident about colour on cans. Multicolour halftones are regularly improving, which has to be for the best as far as the consumer is concerned.

Still one of the major methods of labelling a can is by sticking a paper label round it. Printing on paper is far easier - and cheaper - than printing on metal, but the process does introduce the extra wraparound stage into the manufacturing process, which may reduce any savings that can be made.

Recycling

The metal container industry has, within the last decade, become very concerned with the issue of recycling. First, this helps reduce future production costs; secondly, as consumers and politicians become more conscious of the environment, it has become more than just a question of money. Vending machines that accept and crush empty cans have been installed widely in Scandinavia and are spreading to the rest of Europe.

Collapsible tubes

Another range of products packed in metal are the dentifrice products - essentially toothpastes and creams - contained in collapsible aluminum tubes. The trend here has been towards larger - typically 25mm diameter - packs. In the past 22mm tubes were the most predominant. Perhaps the key developments affecting this sector, however, were the simultaneous introductions of the laminated metal/plastics tube and the upright pump dispenser in 1984.

In various surveys conducted at the time, consumers said that the new packs looked more tidy and attractive than their collapsible metal counterpart and this is a key point as far as designers are concerned. A pack has to look good even when the user throws it away so that he or she is enthusiastic about buying it again. A crumpled, or cracked and torn metal tube is unlikely to endear itself in this way.

New-style packaging

Europe's toothpaste market, worth £70 million in the UK alone, is dominated by major brands from Beecham and Colgate-Palmolive among others. The market is unusual in that it is not seasonal, but advertisements and promotions introducing gels, pumps and new packaging, for example, do affect sales rapidly. However, despite increasing success for the laminated and plastics squeeze tubes, all-metal tubes are still used for a great many products including adhesives, lubricants, contraceptives and skin ointments. The advantages offered by metal as a print medium are also worth bearing in mind and the fact that it can be treated to contain a wide variety of products, almost regardless of chemical content, means that it will almost certainly be a major part of the design portfolio for years to come.

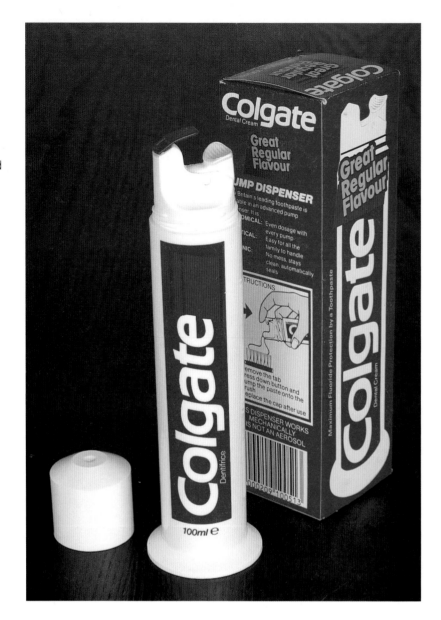

Aerosols

The aerosol is quite different today from its beginnings in the early 1940s. However, many designers still think of the aerosol as a fairly simple cosmetic pack. Nothing could be further from the truth. Artificial snow and de-icers are just two of the many products brought into being by the aerosol can. Many more products owe their convenience to the aerosol, including a wide variety of cosmetics and even pharmaceuticals. In the US, recent launches have included a mousse baby shampoo from Revlon, an antacid product bearing the Maalox brand name and sterile contact lens solutions from Bausch and Lomb. In addition, pack designers are finding that the development of compartmentalized aerosols, often using polyethylene terephthalate (PET), has meant that the manufacture of new products with formulations not compatible before is now possible.

Aerosol manufacture

Aerosols can be made from metals, plastics and from glass, but metal aerosols are the most common, accounting for about 90 per cent of worldwide production. They are made in much the same way as the beverage and food cans described in the previous chapter: the tinplate containers are composed of, usually, three pieces of plate, while the aluminum canisters are usually made in one piece - known as monobloc - by a technique called impact extrusion. The one-piece aluminum aerosol seems to be more popular in continental Europe than in the US, perhaps because a great deal of aluminum is manufactured there. Indeed, there are some interesting cultural differences on the use of aluminum, which are also a result of the manufacturing balance in Europe. Britain, for instance, makes and buys more aluminum cans than its European neighbours, but takes less aluminum foil. The continental Europeans in comparison - particularly the French and the Germans - use foil containers a great deal. The Germans buy much pet food in aluminum foil containers.

The pump-action dispenser for toothpaste is a development that requires careful explanation for use to the customer on the pack carton outer. The strong brand identity of Colgate is maintained by the use of the well-established graphic style of red and white.

Appearance

The main point, though, about one-piece aluminum aerosols is that they look attractive, which is extremely important for the new cosmetics and personal hygiene products, that seem to be attracted to the pack. The smooth canister and the various brushed effects that can be achieved have had a major impact on the fashion-conscious Europeans.

Another advantage, as with drinks cans, is that aluminum canisters do not have the side seam of the traditional tinplate canister, which means designers can create 360 degree designs. In fact, advances in production equipment mean that full-length printing in up to six colours is possible.

Glass aerosols are used mainly for cosmetics, especially perfumes. Small cylindrical aerosols are also used in some medical applications. Production runs of glass aerosols are, however, short and the containers are usually specially designed. Combined with the necessarily slow filling rate of glass containers, glass aerosols are inevitably more expensive than their metal counterparts, but their use for high value items invariably covers this cost.

The crucial aspect of the aerosol is its function as a dispensing pack. There is, of course, more than one way to eject product, depending on the consistency and end use of a canister's contents.

Valves

Perhaps the most important part of the aerosol as a product dispenser is the valve, which the designer should use to define the character of the product it contains. A classic example is a comparison of an aerosol air freshener with a furniture polish. An aerosol for polish should eject large droplets (around 50μ in diameter) that will not blow away in a breeze, while an air freshener should emit a fine, dry mist that will not cloy in an enclosed space. Clearly, the size of the valve opening helps to define how a product is ejected, but other internal factors need to be considered, too, such as the dip or feeder tube. Upright aerosols, such as foam sprays, need a long dip tube, reaching to the bottom of the can, while UHT creams and other food products that are inverted in use need no dip tubes at all.

The all-importance of valves is reflected in the fact that major US manufacturers claim to provide over 3,000 different valves.

The aerosol cap

Most aerosols tend to use a plastics cap that snaps positively over the canister to keep dust out of the valve. Some caps incorporate tamper-evident devices such as rip rings to stop potential buyers testing the product before they get it home. (One chocolate food spray for children was withdrawn because of the enthusiasm with which the intended consumers tested the product in the supermarket.)

Designers of products including air fresheners and fabric conditioners are devising spray-through caps for ease of use. By making the product easier to use - since the cap need not be removed from the container - they actually encourage greater use and therefore further purchases.

The aerosol as food pack

Dairy and food products, too, are receiving greater attention from designers, particularly in the US. Enviro-Spray Systems has devised a quieter multidirectional dispenser for liquids and viscous foods such as mustard, personal hygiene and household products. The Grow-Pak barrier-sealed pouch separates the safe and inert carbon-dioxide propellant from the product to ensure the product is kept pure. The system can be used in aluminum tinplate and plastics aerosols, the company claims, with containers ranging in size from one ounce to two gallons. The company also claims that the container offers an extended shelf life, even for products such as a creosote lacquer, that may only have an intermittent or occasional use.

The Grow-Pak is also used in Metal Box's Trimline aerosol, commissioned by Cuisine et Conserves des Regions de France (CCRF) to pack a range of ketchups, purees and mustard. According to Metal Box, internal lacquering is the key to packaging foods, after a year searching for a compatible lacquer. The external finish was also instrumental in CCRF's choice: a combination of nine metallic and solid colours was required to achieve a striking glossy finish.

Dairy products such as Anchor's UHT cream in an impact extruded aluminum aerosol are increasing in popularity in the UK, but it is a slow process. In the US and continental Europe, in comparison, the concept of foods in aerosols is well-established and successful.

Toiletries and haircare

The success of aerosols continues to hinge upon the toiletries and haircare markets. The worldwide manufacture of aerosols in the late 1980s has reached 8 billion units; Europe on average consumes 3 billion units annually, while the US takes about 2.6 billion. In the UK alone more than 750 million aerosols are filled each year, at a value of £800 million, half of these with toiletry products. Around 17 per cent of aerosols contain deodorants, while a much larger proportion – 20-25 per cent - contain hairsprays. Shaving creams account for over 10 per cent of aerosols filled.

Pharmaceutical and veterinary products products account for a further 10 per cent of the market, but growth is slow. Growth of the haircare sectors, in comparison, has been largely due to the success of mousse styling products, yet another example of a new product in a new pack creating its own new market. When they were launched in 1983, the market was worth £7 million, but by 1987, mousse products accounted for more than £30 million of sales, an increase of more than 400 per cent. Paints and car maintenance products are also an important market.

Aerosols, such as those shown left, look attractive whatever material they are made from, whatever propulsion method they use. This is especially important for the health and haircare markets, where the buyers are very conscious of style and image.

Health and fitness

Other social trends also affect aerosol sales (see also Chapter 1.3). The health and fitness boom is one that has affected many other packaging sectors and it is now contributing to increased aerosol sales. Aerosol suppliers are cashing in on fitness and health by repackaging deodorants, lotions and shampoos in slimline canisters bearing sporty graphics. The slimline container slips easily into the sports kit along with a squash racket, and "two-for-the-price-of-one" or "25 per cent extra free" promotions add to the incentive to buy. The designer working in metals must therefore be as aware of current graphics styles as label designers or pack printers.

The development of taller and thinner aerosols - typically 25-40mm in diameter - also continues. This offers the purchaser a change from the 45mm and 65mm canisters. In addition, because the slimline cans are more portable than their larger cousins, they may appeal to people who have not previously bothered to carry round deodorants or hairsprays before.

Another advantage is that the geometry of the slimline can allows distributors to pack more containers on each pallet, saving transportation space and costs per can. Perhaps more importantly, they allow more facings to show on the retail shelves.

Just the right size for the pocket or purse, this slimline aerosol, moulded in ICI Melinar PET, embodies the flexibility of the aerosol with the lightness of plastics. The stylish but simple type, in minty green, is effective.

Personal hygiene products

For the designer, the key fact to appreciate is that well over 90 per cent of sales in the aerosol market are to women. Many cosmetics manufacturers are therefore keen to develop new and existing product lines to influence male buyers. Elida Gibbs launched an entire range of products under the Sure for Men brand, which instantly proved popular. Sales of the deodorant aerosol alone top 45 million in the UK each year. It is likely that personal hygiene products for men will become a very competitive area for the packaging designers of the 1990s.

Household products

An important sector for the aerosol is household products, accounting for about a third of the world's aerosol production. The major products within this sector are air fresheners, polishes and house plant insecticides. This sector was also the focus for unprecedented competition early in 1987, when the first plastics aerosols began to appear, made from the increasingly ubiquitous PET. One significant aerosol body was the Petasol from Fibrenyle, part of the Canadian-owned Lawson Mardon Group. This was made with Melinar PET from ICI (see Chapter 3.2) and the first commercial user was Johnson Wax.

Reckitt & Colman has also adopted the Petasol, but has renamed it the Super-lite container, plainly making use of the lightweight feature of the container. It is also more tactile, claims the company, and can be decorated without the unsightly seam of tinplate containers.

Environmental considerations

Over the past few years there has been a very significant inter-national concern about the affects on the ozone layer in the atmosphere of chlorofluorocarbon propellants, commonly known as CFCs. From an issue raised by the environmental movements, it is now a major point of concern for consumers and governments.

The surge of interest in the late 1980s, however, was not missed by manufacturers. Initially, they resisted pressure to alter the propulsion techniques their prod-ucts employed, but most are committed now to adopting what are seen as safer methods, realiz-ing that this can be a marketing advantage. In the latter half of the 1980s, small yellow stickers declaring aerosols "ozone friend-ly" began to appear on most canisters in the major outlets. Some designs print the wording on the cap, others incorporate the message into the canisters' labelling.

Propellants

Many manufacturers had already been investigating non-CFC propellants and so the switch did not cause too much added expense. Propellant manufactur-ers from seven countries formed a working party in 1988 to tackle the problem of propellant alterna-tives. Initial participants in the Program for Alternative Fluorocarbons Toxicity Testing (PAFTT) included: AKZO (Netherlands); Allied Signal, Du Pont and Racon (USA); Asahi Glass, Daikan and Showa Denko (Japan); Atochem (France); Hoechst and Kalichemie (Germany); ICI and ISC Chemicals (UK); and Montefluos (Italy).

Whatever the product - from health-care to household - the designers must consider the environment and methods of product propulsion as much as traditional design criteria such as colour or graphics.

European activity

In Europe, Du Pont was particularly active in telling consumers of the potential differences they must expect from the switch to alternative propellants. A lighter bottle and a wetter spray are two side effects of using environment-friendly propellants, and designers must try to use these characteristics in pack designs, rather than fight against them. Wetter sprays, for instance, are ideal for polishes and fabric conditioners.

At a more basic level, having devised a non-CFC aerosol, designers should tell the world by labelling the pack. In many British stores, CFC-free products were labelled as such with the increased awareness among consumers. Tesco, Sainsbury and the Co-op all moved to non-CFC products for own-label goods and Tesco and Sainsbury used labels using the words: "contains no propellant alleged to damage ozone". Sales hardly fell at all. The Co-op, however, said that products that were already CFC-free were not labelled. "It would have been irresponsible from our point of view to start labelling goods with 'good news flashes' purely to entice customers and increase sales."

Conclusion

Aerosol designers today have far more innovations to consider than they did ten or even five years ago. The introduction of slim-line, necked-in and one-piece aluminum aerosols have really ensured that aerosols don't look like stodgy beer cans any more. However, the inherent complexity of the aerosol package - incorporating valve, can and propellant system - means that the aerosol still tends to be fairly expensive to produce but because of their invariable convenience, buyers seem prepared to pay the price.

ICI's UK-based research and development team, for example, is working on even more advanced aerosols. By the beginning of the next century we could all be wearing cotton and polyester clothes sprayed on from a can. At the end of the day we will simply wash them off. Meals from aerosols will also be more exciting, perhaps including protein enriched chicken, beef and fish. A wide range of vegetables, such as mashed potatoes, could even be encapsulated in the aerosol.

It is clear that the possibilities for the spraying pack are almost endless and that among the obvious gimmicks one or two gems will doubtless emerge. How much the ideas evolve from the Research and Development divisions of the world's chemicals manufacturers and how much the pack designers themselves will be responsible remains to be seen.

The demand for effective metal packaging is always under severe cost pressure from competitive materials and methods, and there will always be a strong demand for improved quality and design at minimum increase to cost. Pack design and product development efforts will therefore need to be continued - jointly - on an ever-increasing scale.

The top ten areas of concern for aerosol makers and designers:

- Aerosol inflammability - warehouse and supermarket storage

- The increasing number of US clean air acts

- Use of aerosols indoors - misuse has occasionally caused small explosions or fires

- Birth defects acts - some US states have stringent health and safety acts which ban some products used in some aerosols

- Chlorinated solvents - some attack plastics valves

- Chlorofluorocarbons - propellants and the ozone issue

- Graffiti - controlling the purchase of spray paint products

- Hazardous waste disposal - preventing the explosion of empty cans

- Solvent abuse or "glue" sniffing - sales control

- Increasingly tough product liability insurance - increasing the liability of aerosol manufacturers and designers

Source: Chemical Specialities Manufacturers Association (USA).

4.1

Nutritional and barcode labelling

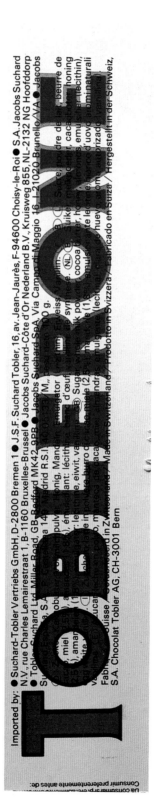

In the nineteenth century, unscrupulous dairies diluted milk, while confectioners added brick dust to chocolate to make production go further. A more dangerous practice was the incorporation of powdered aluminium into bread to make it whiter. In the 1920s and 1930s most packaging and food laws were concerned mainly with ensuring that the buyer was not short changed of goods and that the products they purchased were unadulterated. Legislation requiring manufacturers to list the actual ingredients in foods was first passed in the UK only a few decades ago, in 1953, as an extension of war-time labelling practice.

The 1953 legislation, however, was not far reaching: most foods were bought loose, or as individual items, so it did not have to be. As more and more food factories were established, producing more complex food items, there grew in parallel an intense consumer demand to have more informative labels. As food processing and packaging techniques have changed over the years, so has the requirement for information. And with the strengthening of the European Community, through active participation of countries including Spain and Portugal as well as France and West Germany, the eventual harmonization of individual member states' labelling practices seems certain. There may even be room for some agreement with US Food and Drug Administration guidelines.

How fresh is the produce we buy? Are the drugs on sale over the counter safe? Basic information such as this must be contained on the pack label or wrapping. But as the Western world becomes more concerned with health, even phrases such as "100 per cent natural" have their critics. What does "natural" really mean? A product can contain colourings and preservatives and still contain only natural substances.

The Swiss chocolate Toblerone is immediately identifiable by its shape, but the pack still has to accommodate a large amount of essential information on ingredients in several European languages, reflecting its international market.

The familiar US snack, popcorn, is now available from Golden Valley for the microwave market, so detailed instructions are required on how to prepare it. In addition to all this information there is a table of nutritional data, which helps to sell the snack food to a health conscious consumer.

Pack facings are becoming increasingly crowded with facts and figures about the products they contain. Foods, chemicals and pharmaceuticals, though, are the chief targets of labelling legislation. Most people welcome the changes - food is essential to health, after all, and most of us are interested to see what the drugs we take contain - but many people find it confusing. In the US, the Consumer Network, a research firm specializing in products sold through medium to large retail outlets, claim that: "Consumers believe that labelling is still not clear or understandable on many products." They feel that many product labels are written in a foreign language with no translation offered. Specifying ingredients such as gum acacia and hydrolysed protein does not actually clarify much for most consumers.

Sugar can hide behind different names, for example, so there is a case for using terms understandable by the public.

Small print is not an answer to the amount of information to be presented as it makes customers feel the manufacturer has something to hide. One of them told the Consumer Network: "Small print on the label is a sure sign you should try to buy whatever it is in a different form."

So what must the pack designer display on a label or facing and how best can they get that information across as simply as possible? At a time when the Western world's population is ever more concerned with the safety and nutrition of its diet, labels are the key to avoiding fats, sugar and sodium.

FOLLOW THESE SIMPLE COOKING STEPS AND ENJOY REAL AMERICAN POPCORN—FRESHLY COOKED.

COOK ONLY IN A MICROWAVE OVEN

1 Remove product from carton and cellophane wrapper. Place bag, seam side up, on the oven floor or turntable. **DO NOT PLACE ON A SHELF OR ANY OTHER METAL OBJECT.**

2 Set oven at highest power level: the bag expands as popcorn pops. **IMPORTANT:** Listen for the popping sound! Remove bag when popping slows to 2-3 seconds between pops. DO NOT LEAVE MICROWAVE UNATTENDED DURING POPPING.

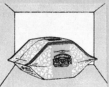

3 Microwave ovens vary in power levels and cooking times will differ from 1½ to 4 minutes. Do not overcook as popcorn may scorch.

4 To open: pull top corners away from each other diagonally, be careful as some steam will escape. **CAUTION:** the bag and popcorn will be hot! For a saltier taste, shake bag before opening.

SAFETY FIRST—DO NOT LET CHILDREN COOK THE PRODUCT

KEEP IN A COOL, DRY PLACE

ADDITIONAL COOKING TIPS INSIDE

Typical Nutritional Information	per 100 g.	per 25 g.
ENERGY	368 kcal/1540 kj	92 kcal/385 kj
FAT	24.0 g	6 g
PROTEIN	5.0 g	1.25 g
CARBOHYDRATES	36.0 g	9 g
DIETARY FIBRE	10.0 g	2.5 g

·GOLDEN VALLEY
MICROWAVE
POPCORN   **100g** ℮

Ingredients: Popping corn, partially hydrogenated vegetable oil, salt, butter flavour, natural colour (annatto).

BEST BEFORE END:
01/90

© 1988

GOLDEN VALLEY
MICROWAVE
FOODS, INC.
197 Knightsbridge, London SW7 1RB
Product of U.S.A.

Legal requirements

Pack labelling law is now extremely complex, but there are a few basic points that can be made. Pre-packed foods must bear:

- the name of the food;

- the weight of the food;

- the ingredients that have been used to make it - if more than one is present;

- how long the product can be stored and under what conditions;

- how the product must be prepared or cooked;

- the name and address of the holding company responsible for the product;

- the country or place of origin may need to be stated, in some instances.

Some of the items on this checklist would not be out of place on other products besides foods, but the food marketplace is the one where legislation has been brought in to help and protect the consumer. Besides the basic facts in the checklist, some manufacturers have decided to provide nutritional information in response to increased consumer awareness. This prompted a demand for more information from other suppliers and now there are comprehensive EC and FDA guidelines about how this information must be presented. For example, because of the increasing concern over fats - chiefly cholesterol - and their alleged relationship with heart disease, most, if not all food labels must now show the total amount of fats present in a food, especially the rich saturated fats. At one time it was thought that milk bottles would have to bear these data, too, spoiling the clean white appearance from the designer's point of view, but this has not happened. Egg cartons, however,

do bear a table of average nutritional data, usually on the inside of the lid or flap.

How to name foods

For the designer involved in the creation of a new food pack, it is as well to discover exactly what one can or cannot call certain types of food. For instance, a squash concentrate must contain at least 25 per cent of the relevant fruit juice. A high juice may contain more. A "fruit flavour yoghurt", on the other hand may never have been near a fruit. The correct description of a fruity yoghurt would be "fruit yoghurt" or "fruit flavoured yoghurt".

Where a product's name gives few clear ideas about what the product is, it must be accompanied by a suitable description. In a leaflet describing food labels and what they mean, UK retailer J. Sainsbury suggests a possible description for a fictional product name - Soufflé Napolitana - which could be sweet or savoury and could contain almost anything. Sainsbury's suggests "cheese soufflé with spaghetti and tomato filling" as an example of one suitable description.

Weight labelling

Foods are manufactured and packed ever more quickly. Coca Cola's flagship filling line in Milton Keynes, UK, fills more than 2000 cans every minute. The French have a taste for pre-packed French salads - automatic weighing machines select and weigh salad stuffs 16 bags at a time, taking only a few seconds to fill each bag. How is it possible to ensure that the right amount gets into each bag? With the advent of high-powered computer-controlled weighing machines, the delivery of all products into containers has grown increasingly reliable. This does not only include foods and pharmaceuticals, but horticultural and chemical products, too.

Until the 1970s, manufacturers had to print what they regarded as the minimum weight or volume likely to be present in any one container. New legislation has meant that only the average weight has to be stated, which some suppliers feel could be bad for customer relations. Before the new legislation came in, buyers always received more than they paid for. However, the tolerances on the weighing machines are very tight and any underweighing is likely to be only slight.

In addition to local national regulations, there are European Community regulations and to show a manufacturer has conformed with these, the symbol "e" is used on labels and pack facings and often embossed on bottles, for instance. This enables foods to be easily imported or exported within the European Community.

Ingredients

All foods composed of more than one ingredient must carry a list of ingredients including added water, if it accounts for more than 5 per cent of the total product weight. Ingredients must be listed in descending weight order with the largest constituent first, but with two exceptions - meats and fish - no data on how much of each is present need be given. However, if a product states that it contains "now even more meat," the percentage of meat it actually contains must be shown.

Some foods are not required to carry a list of ingredients, because their constitution is defined by law. These include butter, most cheeses, fermented creams and milks and alcoholic drinks with more than 1.2 per cent alcohol. Coca or chocolate products are also covered by this ruling.

But what is all this nutritional information for? How can the consumer use it? Most nutritional data relates to a standard quantity, usually 100g (3.5 oz) or 100ml

(3.5 fl oz) of food, but often the amount for the full pack will be displayed, together with occasional symbols representing gluten-free foods or produce suitable for those on a lacto-vegetarian diet, one which may include dairy products.

In California, label warnings are necessary where state-set limits for three so-called reproductive toxins are exceeded. The toxins - alleged to affect babies in the womb - are lead, ethylene oxide and 1,2-dibromo-3-chloro-propane. The state's scientific advisory panel, the Health and Welfare Agency (HWA), states that manufacturers subjecting consumers to more than 0.5 micrograms of lead per exposure, or to more than 20 micrograms of ethylene oxide a day must carry a warning label. They also state that a free telephone number for information or some form of retail notice should be placed beside the foods.

Naturally, the food manufacturers regard this as damaging for business. "This means that anyone shipping lead-soldered cans to California would have to label them especially," according to the National Food Processors Association. However, it adds, no one is sending lead-soldered cans into California. Most are two-piece cans which do not require solder; the others are welded by laser. The lead itself that occurs naturally in the product or in the processing water up to legal limits does not count in 0.5 micrograms.

A range of approaches to the inclusion of nutritional data is shown right, including a version designed to interest the younger consumer.

Legislation along similar lines has been passed in Hawaii, Tennessee and Missouri and if it becomes more widespread pack designers will be forced to do one of three things:

- provide a warranty that pack materials they have used contain nothing on the banned chemical list;

- hold the liability - through the manufacturer - in case of a lawsuit;

- record details of pack composition and chemical formulae to convince customers there is no problem.

An example of data that may appear on the label of a tin of baked beans:

Nutritional information		
Average values:	per 100g	per 850g can
● Energy	315kj (73 kcal)	2677kj (620 kcal)
● Protein	4.9 g	41.7 g
● Carbohydrate	14.0 g	119.0 g
● Fibre	7.3 g	62.1 g
● Fat	0.2 g	1.8 g

Food additives

Because of the diversity of foods and the ever widening techniques of processing them, over the years food additives have become necessary. Additives can be chemicals manufactured in a laboratory or pigments extracted from plants, but all have their function, much tested and then ratified by some body such as the Food and Drug Administration in the US or the Ministry of Agriculture, Fisheries and Food in the UK.

Of these permitted additives, around 300 have been given a number and where the additive has been approved by the EC it has an E in front of it. However, consumers are now wary of *all* additives, as witnessed by E-numbers, or 800-numbers in the US, whatever their composition. This has caused the consumer to attempt to avoid E-numbered foods where possible. However, this trend has itself given rise to a food marketing ploy: the promotion of goods as "additive-free" or as having "no preservatives" has in recent years increased sales of some foods substantially. The designers task of highlighting these trends in on-pack flashes and within the confines of the nutritional label has contributed greatly to the success of this drive.

There are, however, a number of ways of overcoming the negative "E-number" image. Many essential additives have names with less distasteful connotations than an E-number, names which can in fact legally be used on a label, with care. Consider the following whole orange drink:

Ingredients after dilution:
Water, Sugar, Glucose Syrup, Comminuted oranges, Citric acid, Preservatives Sodium benzoate, DSodium metabisulphite, Artifical sweetener (Saccharin), Vitamin C, Flavourings, Colour (Beta-carotene - provides Vitamin A) - not an E-number in sight. Or is there? On this label the designer

and manufacturer have decided to use the chemical names for additives instead of their serial numbers. It is not necessary, either, to number additives that function as acids in foods if their chemical name includes the word acid, as citric acid does, for instance. And Vitamin C has the same chemical structure as the antioxidant E300 (L-ascorbic acid), but here is is being used as a vitamin.

What are additives?

There are a number of basic types of additive, each with a different function. It may be worthwhile for the manufacturers, retailers and pack designers to educate consumers as to the benefits and ask them to weigh these with the disadvantages. Preservatives, for instance, stop microbes from spoiling food or making it unsafe.

New methods of packaging have attempted to reduce the number of preservatives used and so capitalize on the health trend. These techniques include modified or controlled atmosphere packaging and now food irradiation. This has its own public relations problem, but ideas for labelling have been mooted. A logo has been developed and there are agreed forms of words to indicate the level of treatment a product has received.

Antioxidants stop fats and oils from going rancid and destroying essential vitamins. Food irradiation - perhaps surprisingly - is not effective in the fight against rancidity as it sometimes accelerates the process. It has been used in vacuum-packed foods with some degree of success. Emulsifiers and stabilizers, on the other hand, mix or homogenize foods, such as oils and water, and keep them from separating. They are essential for making low-fat spreads.

There are about 60 permitted colourings: the commonest is caramel (E150), made by over-

cooking sugar. Carotenes are red colourings related to vitamin A from carrots and tomatoes. There are also 20 permitted artifical colours, including the azo dyes, one of which, tartrazine, has been linked with allergic reactions in some people, especially children who have become hyperactive. By agreement with the food industry there are no colourings in baby foods. No fresh fruit must be coloured and neither may dried or condensed milks. Tea and coffee cannot be coloured, nor can the flesh of fresh meat, fish or poultry.

There are over 3,000 flavourings, most of them chemically identical to natural flavourings. Honey contains over 200 different substances, each contributing its own special flavour, while apples contain 130 identifiable "flavours."

Barcode labels

The barcode is not as much of a marketing and design intrusion as was first thought. It is increasing in popularity as more and more retailers install electronic scanning systems to improve stock control and speed up the checkout process. Despite being a blot in pack designers' eyes, the barcode can provide useful sales and marketing data. For instance, before its takeover by the Swiss Nestlé group, Rowntree used barcodes to provide better feedback on product promotions and could correlate the sales response to advertising, seasonality or new pack designs. In addition, frozen foods manufacturer Birds Eye has discovered that barcoding gives it new information on the relationship between price and sales volume. And the UK's United Biscuits, which owns the McVities brand, uses barcoding to provide early confirmation of the success - or otherwise - of brand name changes. In short, the future of market research in packaging will be based on accurate and up to date figures due to barcode scanning.

Designers in the Allied Lyons stable, responsible for international brands such as Castlemaine XXXX and Harvey's Bristol Cream, described themselves as very concerned when barcodes were first introduced. They did not initially believe they could safeguard the design of their products, but very soon overcame the visual inhibitions they felt and incorporated barcodes into all their lines quickly and competently.

The UK's Boots chain introduced its first scanning store in 1986, despite concern about the effects of introducing barcodes on pack design. At a 1986 conference to discuss the state of barcoding in Britain, Boots was one of many retailers to suggest that shorter, truncated barcodes were necessary - if against industry standards - because of the small size of cosmetic and pharmaceutical products. For products such as tweezers and nail scissors this problem could be overcome by using blister packs and barcoding the hanging cards instead of the product. The company announced itself firmly behind barcoding in principle even though its designers initially believed that the company could lose years of time in package redesign.

Now designers are finding ever more inventive means of incorporating barcodes into labels. One designer turned the barcode on a bottle of spirits into a veritable reed-bed, complete with bullrushes and wildfowl.

There are more than 30 different barcodes available today, but the most frequently used are Interleaved 2 of 5, Code 39, Codabar and EAN. Interleaved 2 of 5 is among the oldest barcodes available and is suitable for applications ranging from handheld barcode readers to high speed remote scanning. Its one limitation is that only numeric data can be encoded. Code 39 in comparison is useful because digits and letters can be used.

Codabar is the accepted standard for barcodes in the world's blood banks, while European Article Number (EAN), or Universal Product Code (UPC) as it is known in the US, is used in point-of-sale applications in supermarkets, bookstores and hardware outlets, in addition to some non-retail applications.

Designers can select a barcode for an application and help to decide on the relevant technology for generating and scanning the codes by asking a few simple questions:

- will the code require numeric or alphanumeric information to be encoded?
  Solution: Interleaved 2 of 5 or Code 39

- how much information will need to be coded?
  Solution: a short or unlimited code

- does the product require thousands of products with the same code (e.g. in a supermarket), or will each product require a one-off code (e.g. in a pharmacy)?
  Solution: Pre-printed or computer generated labels

- what material is the code to be printed on and will it be covered by shrinkwrapping?
  Solution: this helps to determine the size and density of the code required

- how many times will the code be scanned?
  Solution: choose a printing method that will not be rubbed away

- how will the code be read and at what speed?
  Solution: laser scanner or hand wands can be used with different barcode regimes

Conclusion

If one thing is certain in pack label design, it is that nothing is certain. The speed with which the public interest can change legislation is no legend. It is advisable, therefore, if you operate in label design, to keep up-to-date with regulation-issuing bodies. Lives could be at risk if designers make mistakes.

4.2

Legal requirements

"Buying a pig in a poke" is a proverb which, roughly translated, means you have not seen what you have bought because it is out of sight. The proverb is often used as a warning. Without effective product and packaging legislation, consumers buying goods today could also be said to be buying pigs in pokes. Modern packaging is so effective at protecting and projecting its contents that today's consumers rarely see the goods on offer, except by design, such as the PVC window in a carton, or the cellulose bag. This could tempt unscrupulous suppliers to exaggerate their wares somewhat - in quantity, for instance - if there were no laws to prevent that happening.

Besides preventing pack designs from deceiving the public, more regulations are coming into force to protect the public, too. When US soldiers died from eating tainted beef during the Spanish-American War in 1889-1901, the first Food and Drug Law was passed in 1906. When Tylenol capsules were poisoned in the US a decade ago the resulting deaths brought about new requirements for tamper-evident packaging for over-the-counter drugs. Laws issued by the potent US Food and Drug Administration included guidelines for blister and strip packs, carded packaging and the various seals and bottle closures.

Although the main focus in the US has been on pharmaceutical and drug packaging, the issue is critical for all packagers of food products. It may seem from the preceding chapters that packaging design is being made increasingly easy by the number of new materials available and the variety of processes available for improving them, printing on packaging and filling packs. However, this is not the whole story.

The packaging explosion of recent years, combined with those cases where individuals tamper with foods or pharmaceuticals, for whatever reason, has made it extremely important for the designer to understand how goods must be protected under the law, as well as in practice. But the two are not disconnected. It is as important for modern designers to understand how legislation affects packaging design as it is to understand, for instance, how those designs can be printed. The main thing to realize is that "safety" can be designed into a pack as easily as colour.

Designing for safety

The way in which the design of safety is organized varies from company to company and from individual to individual. But one approach, used in the electronics industry, is to stage six separate design reviews where the following 12 points are considered:

- reliability
- performance
- maintenance
- manufacture
- product test
- interchangeability
- installation
- simplicity
- safety
- ergonomics
- appearance
- cost and value

The first review, known as the design concept review, is held before the development work even begins, while the second deals with the design approach and the specification of materials and processes. The subsequent four reviews cover the basic design itself; experimental data and value analysis; the planning of product manufacture, and a final design review. Not all packaging sectors need be as stringent as the electronics industry, but those that do use or implement such safety reviews can only instill confidence in the retailers and, ultimately, the end-user on whom the whole design and manufacturing process depends.

Legal liability

A general checklist of the areas in which pack designers must consider legal liability includes the packaging of chemicals, where chemical burns might occur if a pack leaks. The products, liquid or powder, may be inflammable, or react with the atmosphere to emit gases or heat. The designer must think of these possibilities in advance. The packaging of light electrical goods must protect the user from possible electric shocks and not be dented in transit at pinch or crush points. Bottles and containers of chemicals may explode or implode depending on atmospheric pressure. This also extends to aerosols, drums of steel or plastics, barrels and kegs of beer and sparkling wines.

In addition, there are problems of end use. Textiles and empty aerosols may well be inflammable or explode if left near heat sources, so adequate warnings must be given on the pack outer. Products containing water should also be advised on temperature, especially if there is a danger of freezing - as water expands a pack could rupture if stored at a cold site.

Product description

Besides health and safety warnings, advertising standards rules must be followed. If a battery, say, is defined as re-usable, any process that is necessary to recharge it must be adequately described, on the pack as well as in any advertising material.

The problems of packaging radioactive goods offers an extremely difficult task to the designer involved mainly in consumer work, as all Western countries have extremely strict rules to follow. If faced with a project on radioactive goods, for hospital work, for example, the designer must get expert advice. It is not enough simply to read the available reports and standards material.

Needless to say, items such as razor blades, knives, garden implements and tools all require effective, safe packaging. But the designer must also think how to protect consumers against shrapnel and flying objects from bottles, canisters or aerosols, where these are contained within cartons or tins.

Stability can also be a problem - tall, thin packs such as half-empty bottles could be described as unstable. If they fall over, will the contents be dangerous? Will the container explode?

Two further points should be borne in mind, though, when considering packaging design and potential dangers that might arise: severity and frequency. If a danger is identified, will it be minor, such as a leak in a carton, for instance, or will it possibly cause blindness, as an exploding pressurized container such as a bottle of soft drink might? Again, what is the chance of the hazard actually occurring? Is it remote, or likely, given the pack's traditional distribution and use? Designers cannot afford to spend all their time worrying about product liability, but they must give enough care and attention or be found liable of criminal negligence.

European legislation

In Europe, there are two main sources of legal standards: governmental legislation and European Community directives. Unlike many standards and codes of practice, which are advisory in nature, these must be followed. UK legislation, for example, often evolves a British Standard - or a British Standard is turned into legislation - and these are not then simple guidelines for the designer, they are mandatory.

Standards based on UK specifications have been adopted by other European countries, most notably in medical or surgical packaging. Germany has its DIN standards; the Netherlands has OVON. The medical packaging companies themselves are governed by British Standards, which ensure that sterilized paper packs, for instance, are of the highest quality. Physical properties specified by the standard include substance or weight, tensile strength, tear surface absorbency and characteristics called dry and wet burst.

With many products, packaging warehousing and distribution are vital for product and personal safety, and so in many areas packaging is covered by both British Standards and EC Directives. A spokesman for Welton Packaging's medical division said: "We have to be able to trace a bag not only backwards to the source of the paper supply, but forwards - how many did we make, where did they go and so forth. This demonstrates that if there was a problem, with the design, say, we could recall them from the marketplace."

In some instances, individual governments resist new forms of packaging if they believe the new form will damage an important existing industry. It was only after five years of trials that the Italian Ministries of Agriculture and Public Health gave the go-ahead for wine to be sold in cartons and

Beer and wine labelling is strictly governed by legislation in the producing countries and now the European Community. This traditional beer from Cornwall, UK shows the amount of alcohol by its 'original gravity', whereas the wine gives alcohol content as a percentage. This German wine label carries an official testing number (amtliche Prufungsnummer) with details of where it was produced and bottled.

plastics bottles. There were fears for health, but the major concern was it seems over the threat to glass. The decision did not, however, apply to wines that were meant to age in the wood or to DOC (appellation controlée) vintages. In addition, say the Ministries, the containers must bear an expiry date: nine months for cartons and six months for plastics containers. The results of tests on bag-in-boxes, cans and demijohn-sized bottles in polyethylene were due to be released by the end of 1989.

Following reports in the mid-1980s that the ClingFilm type of PVC food wrappings constituted a possible health hazard, protracted discussions between various European bodies resulted in draft legislation being issued with significant implications for packaging designers and manufacturers. Under the new regulations, the responsibility for testing and ensuring that carcinogenic chemicals within plastics do not pass into food - beyond preset limits - falls squarely on finished packaging designers and producers, as well as on product

packers. The additives in question are the plasticizers which give PVC its cling.

The Packaging Industries Research Association in the UK, undertaking research on behalf of the European Community, believes it must emphasize the importance and necessity of its work to industry. "When the EC legislation comes into effect it will not be sufficient for retailers or packaging producers to say their suppliers have assured them materials are satisfactory. They will probably be required to present evidence that they have undertaken the necessary precautions to ensure that packaging has been tested and meets the requirements of the regulations."

Clearly the EC legislation is important - without it each government will adopt a different policy. The West German health authority withdrew its proposed restriction on the use of plasticized PVC when it heard that the EC was introducing legislation.

US legislation

In the US, designers must check Federal Acts and Regulations in conjunction with local state requirements. The two can sometimes be at odds with one another. Certain products or labels cleared by a state, for instance, need to be approved by one or more federal agencies before they can be used. States that export goods to other states must meet the receiving state's regulations. California, Missouri, Hawaii and Tennessee are particularly strict.

In general, meat and poultry labels must be cleared by the Department of Agriculture's Food Safety and Inspection Service (FSIS). In addition, manufacturers of food packaging must ensure that their materials meet criteria defined by the Federal Food, Drug and Cosmetics Act (FFDCA).

As regards foods and food additives, the Food Additives Amendment to the FFDCA places the burden of proof on the manufacturer, but the manufacturer's responsiblity also includes ensuring that packaging materials conform to regulations.

Among the most significant exceptions in the US is acryclonitrile (AN), which was originally developed as a bottle-blowing material for carbonated drinks. However, there are allegations concerning the carcinogenic nature of the material as a result of tests on some animals and it is now banned from drinks packaging in the US. The ban seems unlikely to be extended to the European Community, however, unless specific research is carried out on toxicity relating to humans. PVC for spirits bottles is also banned, as are fluorocarbon propellants for aerosols.

Some of the most recent changes to US law were made by the National Fire Protection Association (NFPA) for manufacturers of inflammable liquids, which now face significant changes in the way they store and pack products in plastics containers. Tests carried out by Factory Mutual Research Corporation showed that plastics jerricans weakened by heat could actually squirt boiling paint thinner, which would then catch fire and engulf a pallet in flames.

Storage of these liquids - including chemicals, paints, petroleum products and solvents - will require separate inside rooms or warehouses specifically equipped for storing inflammable liquids. There is some confusion in the US over the revisions to the fire code and more revisions were due in 1990, so pack designers must investigate the code thoroughly and take the most effective action if involved in plastics containers for inflammable liquids.

Tamper evidence is a problem the US is particularly concerned with following the Tylenol killings in the mid-1980s, where pharmaceutical capsules were laced with cyanide. Designers should welcome, then, the Food and Drug Administration's latest tamper-evident regulations. Measures no longer considered acceptable by the FDA include Cellophane wraps that overlap and are not heat sealed. Tinted wraps were dropped, together with wet cellulose shrinkbands, which can be refitted. Tape seals that offer no visible evidence of tampering were also criticized as were glue-sealed cartons.

The FDA urges that bands, wraps and pouches remaining on the acceptable list be printed with a distinctive design that cannot be matched easily once removed. It would prefer a different design for each product use. It has also requested changes for blister and bubble packs. These should be so tightly sealed that the two cannot readily be separated or easily replaced without leaving evidence. The administration also expresses a preference for tear strips that signal prior opening, breakaway caps and tubes that must be punctured to get at the product.

Packaging law

The difficulties facing the packaging designer are exacerbated by the fact that there is no distinct collection of laws that may conveniently be described as the laws of packaging. Packaging designers need to consider not only sale of goods acts, but also trade descriptions laws, transport legislation, weights and measures acts, food and drugs law and the many poisons and medicine statutes that exist. In addition, there are laws and regulations that relate more directly to the design process: there are copyright, design and patent laws that help protect against negligence, passing off and fraud, amongst other things. Yet in a sense this bewildering variety of laws and regulations relating to packaging corroborates the thread running right through this book, that packaging impinges on all aspects of our lives, both commercially and as individuals. The task of the designer is an extremely important one, then, in tying together all these threads.

Perhaps the most important requirements of packaging law are to ease administration. There must be set requirements for food packaging to safeguard every consumer, just as there must be specific requirements for packaging dangerous goods. If designs follow the set guidelines then their ability to pass into society safely and rapidly are guaranteed. Of course, laws must protect the public, and by following the guidelines of neighbour countries, or the countries into which goods are likely to be exported, we protect the public. Packaging law also protects the designer. Designers must do everything in their power to protect the public from injury and to comply with the relevant administration. It is only professional that they should do so.

Food and drugs legislation

Because of the importance of food and drugs, every country has extensive food and drugs laws to ensure that no foods that endanger human health shall be sold for human consumption. Contravening the acts can result in hefty fines or imprisonment. The acts also prevent misleading claims being made about foods, either in on-pack displays or in general supportive advertising. Various laws are regularly added, depending on governmental decisions and current social trends, such as the concern with health and nutrition discussed in the previous chapter.

Food labelling has become an extremely emotive issue and clear, unequivocal descriptions of foods are required, together with lists of ingredients and the manufacturers' name and address. Labelling laws even specify how large the printed type must be on food packs and it is construed as an offence if the type is too small or insignificantly placed. These suggest the supplier is misleading the consumer as to the nature, substance or quality of the food, or of any ingredients. As far as hygiene is concerned there is little specific legislation related to packaging except where provision is made that no materials should come into contact with food they are likely to contaminate.

Dangerous substances

It is generally held in law that any person or organization that allows a dangerous substance to escape is liable for the consequences. The laws cover poisons, pharmaceuticals, medicines, drugs, and dangerous chemicals. The major precautions prescribed by law include the careful handling, packing and labelling of these goods. Failure to comply with the regulations can lead to large fines and, in extreme cases, imprisonment.

Besides any legal ramifications there are pure packaging considerations to look for: bottles for poisons must in general be fluted vertically with ribs or grooves so that they may be recognized by touch. Some plastics moulders supply ribbed bottles of household bleach, although this is a borderline poison. Most moulders - and some glass blowers - now include an embossed message in braille for the blind and poor sighted.

Dangerous businesses such as match making and slaughtering have little direct effect on packaging design, but there are occasionally implications - of inflammability, for instance, in match making - that may have to be contended with. Of more direct concern is the transport of petroleum and other inflammable or corrosive liquids. The way these liquids can be packed or stored, and the very nature or construction of their containers is very strictly laid down in a series of laws for every Western country.

Transporting dangerous goods

Most regulations for transporting dangerous goods are reasonable and straightforward and state that packs shall not leak. The United Nations Committee of Experts, recognizing the impossibility of implementing a worldwide code for hazardous goods, has nevertheless drawn up a regulatory framework which individual countries and states can revise as necessary to suit their particular requirements. The UN's object is eventually to move every country towards a common ground for legislation.

There are some peculiarities with this system, though. The producer/exporter taking goods through a series of countries must comply with the regulations in all the countries through which the goods may pass. This is not simply a case of complying with the most stringent of regulations, as each country may have different classifications of dangerous goods and there may be no common denominator.

A number of important steps have been taken, though, in the drawing up of regulations concerning road and rail travel. There is a European-wide agreement for the carriage of dangerous or hazardous goods by road, known as ADR, which provides a common classification. The ADR imposes specific packaging standards and labelling regulations. The regulations also cover the transport of goods by sea. There is a separate agreement covering the international carriage of hazardous goods by rail and this is known as the RID. It divides dangerous goods into seven classes, depending on the potential severity of an accident. Other provisions are made outlining which goods must be kept separate and how they should be labelled and documented.

Protecting the designer

Protecting the designer, while much less important to society, is also relevant. There are separate statutes relating to pack design, including copyright, design, technical patents and trademarks. In each case the object is to protect the designer against the unauthorized use of his or her work. Copyright laws, meant to protect the original investment of time made in the creation of some work, have been extended considerably since they were first introduced, and now include diagrams, charts, plans, paintings, drawings and photographs.

Copyright

The copyright in most cases belongs to the company that employs the designers, rather than the designers themselves. Copyright does not need to be registered with any organization and is automatically assigned to the creator of the work, unless it has been commissioned. Copyright usually lasts for a period of 50 years from the death of the artist or from the work's commissioning. The actual details vary from medium to medium, but there are certain exceptions in pack design. Advertising slogans and phrases are deemed generally not to be covered as most are not thought to be original. So clearly no copyright will attach to a carton or paper bag as copyright relates to original work only. But a design for a delicate plastics moulded in the shape of a milk churn, or an innovative aerosol, is more likely to succeed in its application. However, only in extreme cases is shape covered, no matter how artistic and original. This was highlighted most recently by Coca Cola Corporation's attempts to copyright the famous shape of its bottle. It failed in the attempt, which should leave pack designers in little doubt over their position, at least for the time being.

Patents

Copyright also exists in industrial designs, but these must be registered, in the UK with the Patent Office. Fees are payable for the registration of industrial designs. Patents themselves are a strong form of protection, although more relevant to the technical side of pack design. Again, fees are payable for the registration of a patent. The periods of registration are usually much shorter than copyright periods, lasting around 15 years, although they can be even shorter.

Trade marks

Trade marks are symbols, which may include words, a graphic device or both, which the designer uses to distinguish the owners goods from any other. Trademark rights are less powerful than copyright, or patent laws, in that they need not be registered and the regular use of the mark may, in common law, establish a statutory right to use the mark exclusively. However, marks can be registered and the more distinctive and original a mark is, the more likely it is to be granted registration.

There are certain criteria for registering trademarks, which may be composed of one or all of the following:

- an invented word, or phrase;

- the name of a company, individual or firm represented in a special or distinctive way;

- a word or phrase having no direct reference to the character or quality of the goods and not being a geographical name or surname;

- any other distinctive mark.

Once a trademark has been successfully used for a period of seven years it may then be renewed for periods of about 14 years indefinitely. The laws differ in different countries.

International copyright

With the ever increasing scale of international trade, the question of international copyright has become an issue, and the subject can be very complex. There is, however, a degree of flexibility, as publication in one country or state is often deemed to have been published in other countries. This will clearly be the case in Europe by 1992, although it occurs in some measure today. In the US similar conditions exist. However, there are exceptions, usually relating to photographs or illustrations. In an attempt to keep production costs down, pack designers sometimes purchase only the UK or US rights to a photograph or illustration. The main reasoning being that it seems unnecessary to pay the extra expense for international rights when only the home market is being supplied. But when the time comes to export the product, the designer must ensure the rights to the work are obtained before continuing with the export.

In many cases the designer has long since passed on the work and this responsibility falls to the production manager, but it is a point designers should bring out at the design stage.

A similar activity to breaching copyright law is known as "passing off," where a supplier attempts to simulate some well-known brand with a similar pack. This can occur by using similar colours or graphics in the design, or by adopting the same shape. The goods are usually inferior. No action for copyright can be taken by the injured party because the pack is never usually the same, but the bogus trader has certainly misled consumers. If this happens the honest trader's response is to take a civil action for passing off against the imitator. The practice is usually stopped and damages can be brought against the defendant, where relevant.

The history of packaging

Year	Paper and paper products	Glass	Metal	Plastics
8000BC	Woven grasses, soon replaced by cloths	Clay pottery and crude glassware		
1550BC	Poultry wrapped in palm leaves to protect against contamination	Bottle making is an important industry in Egypt		
200BC	Developed by Chinese from mulberry bark			
Greek and Roman times	Wooden chests, kegs and barrels	Bottles for perfumes, jars; earthenware urns and bottles		
750AD	Paper-making reaches Middle East; from there reaches Italy, Germany			
868	First evidence of printing - from the Chinese			
1200	Paper-making reaches Spain; from there reaches France, UK in 1310		Tinplated iron developed in Bohemia	
1500	The art of labelling is created; Jute sacks widespread			
1550s	Oldest surviving printed wrapper from Andreas Bernhardt, Germany			
1700	Paper-making reaches USA	Champagne invented by Dom Perignon - only possible because of strong bottles and tight-fitting corks		
1800		Jacob Schweppe started business in Bristol, England as a maker of mineral water - Schweppe's; Janet Keiller of Dundee, Scotland, sold the first orange marmalade in wide-mouthed glass jars	Handmade soldered tinplate canisters in use for dry foods	
1810			Peter Durand devises cylindrical sealed container - the can	
1825	Druggists in the UK adopt regulations for the labelling of poisons		Aluminium isolated from ore	
1841	Paper boxes cut and creased by hand. Screw cap patented 1856		Collapsible tubes first used for artists' paints	
1890s	Printed paperboard cartons appear. Crown cap patented 1892	The first milk bottle appears; Scotch whisky appears in London and is exported. James Buchanan's House of Lords brand soon becomes known as Black & White because of its label; Coca Cola appears in bottles - Pepsi Cola soon follows	Toothpaste invented - starts to appear in collapsible tubes	
1900s	Uneeda biscuit package ousts the tinplate biscuit barrel. MW Kellogg launches cereal packet	Mayonnaise is bottled in 1907	Aluminium covers made for Mason jars	
1905	Composite paperboard cans appear - some spirally wound. Fibre drums for cheese also designed		Steel barrels are designed to carry oil for Standard Oil (now Exxon); they replace wooden barrels. Oxo design - white lettering on red tinplate container - first appears in early 1900s	
1909	Wirebound crates appear for bulk packaging			Cellulose acetate developed for photographic use. First film generating machinery developed in Switzerland 1911

Year	Paper and paper products	Glass	Metal	Plastics
1900-30		Perfume bottles become more adventurous	Foil wrapper used (1913) for US Life Savers candy bar	
1924		The UK's United Dairies becomes Britain's first dairy to switch to bottles for its milk deliveries		Du Pont manufactures first Cellophane in New York
1927				PVC available as a commercial product
				Expensive plastics caps are used on luxury items
				Polyester - a British discovery - was bought by Du Pont and licensed to ICI for European distribution. This lead to the development of polyethylene terephthalate 12 years later
1928		The US baby food industry starts packing products in glass jars		
1933				ICI develops polyethylene; Germany develops polystyrene
1938				Du Pont introduces Nylon
1940			Aerosol devised as DDT spray	A type of polyethylene used to pack Mepacrine tablets in WWII
				Better production techniques developed in 1946
				First tubular bag blown in 1949
1947				Squeezy bottle designed for Stopette deodorant
1948				First shrink-wrapped product: turkeys for deep freeze storage
1950s			First aluminium foil containers	High-density PE developed in the UK and the USA, by Phillips Petroleum and Standard Oil (Exxon);
				Polycarbonates developed by General Electric and Bayer (West Germany)
1959			The aluminium can is designed	Polypropylene, developed in Italy, first appears as a film
1960s				LDPE used for heavy duty sacks for fertilisers
1973				Stretch wrapping introduced in Sweden
1977		Glass starts to be used only for high-value products		PET becomes widespread as bottle material for carbonated drinks
1980s			Continued down-gauging of tinplate containers; moves to design only two piece cans; resurgence of interest in tinplate as a nostalgic medium	PET used for foods and hot-fill products such as jams
				High-barrier, multilayer containers increasingly used
				Guy la Roche uses PET for perfumes
1990s	Increasing use as designers aim to cash in on the 'green' revolution	Glass regains more attention as a recyclable pack medium		More designs incorporate biodegradable plastics

Further reading

Books

Abbott H (1980)
Safe enough to sell? Design and product liability
The Design Council, London

Boustead I and Hancock G F (1981)
Energy and packaging
Ellis Horwood, Chichester

Briston J H (1983)
Plastics films (second edition)
George Godwin,Edinburgh/ Plastics & Rubber Institute, London

British Plastics Federation (1980)
Users guide to blow-moulded plastics containers
BPF, London

Griffin R C, Sacharow S and Brody A L (1985)
Principles of package development (second edition)
AVI Publishing Company, Westport, Connecticut

Hanlon J F (1984)
Handbook of package engineering (second edition)
McGraw Hill Book Company, New York

Kahn M R and Rahim E (1985)
Corrugated board and box production
Scottish Academic Press, Edinburgh

Moody B (1977)
Packaging in glass (revised edition)
Hutchinson Benham, London

Morgan E (1985)
Tinplate and modern canmaking technology
Pergamon Press, Oxford

Paine F A, ed (1985)
Fundamentals of packaging (7th edition)
Blackie & Son, London/Institute of Packaging, Melton Mowbray

Pilditch J (1973)
The silent salesman (second edition)
Business Books, London

Urbain W M (1986)
Food irradiation
Academic Press, Orlando

Webb T and Lang T (1987)
Food irradiation: the facts
Thorsons, London

Market research literature

Covell P, Sonsino S, Sowerby P and Everest E (1987)
Annual statistical summary 1986
Institute of Packaging, Melton Mowbray/ Online Business Publishing, London

Fairley M (1985)
Bar coding technology: an equipment review
Labels & Labelling Data and Consultancy Services, Potters Bar, UK

Henley Centre (1982)
Manufacturing and retailing in the 1980s: a zero sum game?
Food Manufacturers Federation, London/ Nielsen Marketing Research, London

ICC (1985)
Paper and board packaging
An ICC Business Ratio Report
ICC Publications, London

ICC (1985)
Plastics packaging
An ICC Business Ratio Report
ICC Publications, London

Keynote (1985)
Packaging (Metals and aerosols)
Keynote Publications, London

Keynote (1985)
Packaging (Paper and board)
Keynote Publications, London

Keynote (1985)
Packaging (Glass)
Keynote Publications, London

Keynote (1985)
Packaging (Plastics)
Keynote Publications, London

Martin W H and Mason S (1982)
Leisure & Work:
the choices for 1991 and 2001
Leisure Consultants, Sudbury, Suffolk

Mills R (1985)
UK packaging industries: facts and data
Pira, Leatherhead, Surrey

Mills R and Lushington R (1985)
The UK packaging industry: its structure, suppliers and future outlook
Economist Intelligence Unit, London

Skyme M D (1981)
The soft drinks market
M D S Services (Publications), New Quay, Wales

Journals

Australia
Australian Packaging
PO Box 204, Strawberry Hills,
New South Wales 2010, Australia

Austria
Austropack
Verlag A Schendl, Karlgasse 15,
1040 Vienna 4, Postfach 29 Austria

Denmark
In Pak
3 Kongsatrupvcj,
Box 15Dk-4390, Vipperod, Denmark

East Germany
Die Verpackung
VZB Fachbuchverlag Leipzig, 7031 Leipzig,
East Germany

France
Emballage
26 rue de Fbg Poissoniere, 75010 Paris,
France

Embouteillage
7 rue de la Boetia, 75008 Paris, France

Italy
Imballagio
Etas Kompass Periodic Techni, Via Nuova
Rivoltana, 20090 Limito, Milano, Italy

Rassegna dell'Imballagio
20156 Milan, Via Casella 16, Italy

Techniche dell'Imballagio
Vialle Monza 106, 20127 Milano, Italy

Netherlands
Missets Pakblad
Postbus 4, 7000 BA Dostinchem, Netherlands

Verpakken
Postbus 81152, 3009 GD Rotterdam,
Netherlands

Verpakkings Management Postbus 247,
1700 Ae Heerhugoward, Netherlands

Norway
Emballering
Postboks 855, N-3001 Deammen, Norway

Spain
IDE
Breton de los Herreros 57, Madrid 3, Spain

Sweden
Nord-Emballage
Box 25, 16211 Vallingby-1, Sweden

Packmarkenden
PO Box 601, S-25106 Kelsingborg, Sweden

Switzerland
Die Verpackung
Forster Verlag, Alte Landstrasse 43,
8700 Kusnacht/ZH, Zurich, Switzerland

Tara Verlac
8032 Zurich, Beustweg 12, Switzerland

UK
Design Week
Centaur Publications, London, W1A 2HG

Marketing
30 Lancaster Gate, London, WC1A

Packaging Today
Angel Publishing, 361 City Road,
London, NW1A

Packaging Week
Benn Publications,
Sovereign House, Tonbridge, Kent, UK

USA
Food & Drug Packaging
7500 Old Oak Boulevard,
Middlesburg Height, Ohio 44130, USA

Packaging
Cahners Plaza, 1350 East Touhy Avenue,
PO Box 5080, Des Plaines,
Illinois 60018, USA

Packaging Digest
400 North Michigan Avenue, Chicago,
Illinois 60611, USA

West Germany
Pack Report
Schumanstrasse 27, 6000 Frankfurt am Maine,
Postfach 7625, West Germany

Packung und Transport Kreuzstrasse 21,
Postfach 1102, 4000 Dusseldorf 1,
West Germany

Verpackung Berater
P Kepler Verlag, Industriestrasse 2,
6056 Hausenstamm, West Germany

Verpackung Rundschau
P Kepler Verlag, Industriestrasse 2,
6056 Hausenstamm, West Germany

Glossary

adhesive A general term covering cements, glues, pastes and thermoplastic adhesives.

aerosol There are three basic methods of discharging a material through a valve from a container using pressure: true aerosols, sprays - such as paints and waxes - and foam products, such as shaving creams and hair preparations.

aerosol containers These may be reusable, throwaway or single-use - usually made from tinplate or aluminum - and glass bottles with a neck finish to take the aerosol valve.

aluminium foil A solid sheet traditionally rolled to a thickness of less than 0.006in (0.015cm). This can be laminated on to other materials, such as polyethylene, to provide speciality high-barrier films.

B-flute See flute.

bag A preformed container made from flexible materials, now usually paper or plastics. They are generally enclosed on three sides, and the remaining side may be sealed after filling. Bags can be made in several plies, using different materials coextruded or laminated together.

barrier materials A material designed to withstand the penetration of water, or water vapour, oils and certain gases.

blank A piece of material from which a container or part of a container will be made by further working.

bleed To print an area beyond the cut edge or score, so that the design is cut off or, as in folding cartons, so that the design is folded under an outside flap.

blister A small, localized area free from adhesive, used to great effect in the manufacture of the now-widespread PVC blister packs. Also known as strip packaging when small articles such as tablets are packed individually in a continuous strip.

blocking An adhesion between touching layers of material under moderate pressure or temperature. Often used to decorate glass or ceramics. Occasionally occurs unintentionally in storage or use.

board A heavyweight sheet of paper or other fibre substance usually thicker than 0.006in (0.015cm), although the distinction between board and paper is not specific. Boards for packaging may be made from kraft mixed with waste papers, manilas, fibre, and newsprint among other materials. Boards may also be laminated or coextruded with other materials such as plastics and foils.

body The main part of a container, generally the largest part including the sides.

BSI British Standards Institute. A near-equivalent to the US American Standards Association.

CAD Computer-aided design, the technique of using high-resolution computer graphics systems to help design packaging. Also known as computer-aided packaging design (CAPD).

cardboard A colloquial term, not generally used in the specification of container materials - use paperboard, fiberboard or board as necessary.

carton A form of pack made from bending grades of paperboard.

case A non-specific term for a shipping container.

Cellophane A transparent film made from regenerated cellulose. It is inherently greaseproof and with suitable coatings may be made moistureproof and heat sealable.

cellulose A carbohydrate constituent of plant cell walls.

cellulose acetate A thermoplastic material made by the treatment of cellulose with acids. It is usually extruded into films and may be extruded or molded into containers.

CFCs Chlorofluorocarbons, the propellants used in aerosols until the late 1980s, when research showed they might be responsible for the depletion of ozone in the atmosphere.

closure A sealing device or covering that attaches to a container and retains the contents, preventing contamination.

coating A covering or layer of a substance deposited in fluid form and dried on to the surface of a material or product to inbue some characteristic, such as barrier properties.

collapsible tube A cylindrical container of thin, flexible materials with integral shoulder and neck. Now almost exclusively made from plastic-foil laminates, but also from tin-lead alloys and aluminium among other materials.

corrugated board Board that has passed through a corrugating machine. Single-face corrugated board is made from one ply of liner attached to one ply of corrugated medium. Double-face or single-wall uses two plies of liner material sandwiching the corrugation. Double wall has three plies of liner and two corrugations.

crimp To fold in, squeeze, or tighten by a series of corrugations so as to hold one part against another. Milk bottle tops are one example, but the technique also applies to the tops or bottoms of metal cans.

crown A metal closure lined with cork, used for narrow necked carbonated beverage bottles.

deadfold A hand or machine-made fold which remains in place without sealing. Usually refers to soft foils.

die-stamping The process of reproducing a design, figures or lettering from engraved, usually copper or steel, printing plates.

enamel A vitreous, paint-like material used to decorate or protect. It is usually baked on to the substrate.

etch To treat a material with acid, leaving part of the material in relief to form a design.

extrusion The process of forcing molten materials through an aperture or die to form continuous lengths of sheeting, film, rods or tubes. The material is then immediately cooled to make it retain the new shape. Impact extrusion - as used in the manufacture of one-piece aluminum aerosols - is the process where a die or mould is charged with a pellet or disc of metal and forced by impact to conform to the shape of the die.

fiberboard Fiber sheets produced or laminated to a thickness providing stiffness. Fiberboard for the manufacture of cartons may be corrugated or solidboard.

film Unsupported material, basically organic and non-fibrous, that is flexible and does not usually exceed 0.01in (0.025cm) thick. Film thicker than 0.01in (0.025cm) is technically termed sheet or sheeting.

finish The quality of a surface - including colour, brightness and texture. In paper and paperboard specifications finish has a special meaning: it is a measure of smoothness ranging from 1 to 4. Number 4 is smoothest and most dense.

flexible packaging Packaging that uses flexible materials such as foils, films, paper or sheeting to form the container.

flexography A rotary letterpress printing method that uses rubber plates and fast-drying transparent inks. It is a relatively cheap process and so is often used for short-run, low-volume products, or in the execution of 'flashes' or special promotions.

flute A rib or corrugation on a surface, usually classified as shown in the diagram on page 00.

foil Unsupported thin metal membrane less than 0.006in (0.015cm) thick. Above 0.006in (0.015cm) the metal is called a sheet. In many European countries the word foil - from the French, feuille - means any thin material, including Cellophane and plastics films as well as thin metals.

gauge A method of indicating the thickness of a film or metal in which the numerical prefix is related to the thickness of the material. In films, for instance, 88-gauge is equal to 0.00088in (0.0022cm).

glassine A smooth, dense, transparent or translucent paper made mainly from chemical wood pulps. Made in white and various colours, it can be considered a high-quality greaseproof paper for wrapping foodstuffs, tobacco, chemicals, and metal parts such as sharp blades.

gravure printing An intaglio process using tiny engraved wells. Deeply etched wells carry more ink than a raised surface can so offer darker, higher quality printing. The process is more expensive than flexography and so is only used on long-runs, where the cost becomes economical.

halftone A printing plate used to reproduce a photograph or other 'continuous tone' design by dots of varying sizes. The term also refers to the impression that results from the printing process.

heat seal A means of uniting two or more surfaces by fusing them or their coatings together.

hologram Three-dimensional image reproduced on foils and films - recently becoming more widespread on cartons and magazine or book covers.

imprint The informative legend printed on a container during the manufacturing process. This should contain the maker's name, container capacity, material quality, freight classification, and any other information depending on the regulations that apply in the country in which the product is sold.

jar A rigid, flat bottomed vessel with a wide mouth.

lacquer A type of coating applied in liquid form, which dries by evaporation.

lithography A printing process using flexible metal plates whose printing surfaces are partly water and partly oil-repellent. The process is especially adapted for fine halftone colour effects on a variety of papers and boards as well as on metals.

machine glazed (MG) A treatment for paper and board.

metallized film A film or paper substrate coated with vapourised molecules of metal in a vacuum chamber is said to have been metallised. This improves the barrier properties of the material and is cheaper than using foil.

modified atmosphere packaging (MAP) Generally, food packaging, in a container where the normal air has been replaced by a special gas mixture of ordinary gases that extends the shelf life of the food naturally.

multiwall Having more than one wall or ply. In the case of bags, this tends to mean more than two walls or plies; two-ply constructions are usually called double-wall or duplex.

neck The narrow upperpart of a container, between the shoulder and the opening. Hence 'neck in', to form a portion of a container to a smaller size than the main part or body. This is used to great effect in the production of beverage cans, where smaller aluminum ends are used, thus saving on the costly metal.

opacity The resistance of a material or body to the transmission of light.

oriented film A film such as polypropylene in which the molecular structure is aligned mechanically in one or more directions to gain characteristics such as strength or added shrinkage.

overprint The result of printing one layer over another, such as a varnish printed over an ink film to protect it.

pack To put material or goods into a container for storage or transportation. Also, a complete range of one product in a container, e.g. the toy pack.

package One unit of a product, wrapped or sealed in a sheath or container. A container in which a product is packed.

pallet A low, portable platform, usually of wood, but increasingly now of metal, fiberboard or plastics.

paper A general term referring to matted sheets of vegetable fiber. Synthetic papers from mineral, animal, or synthetic materials can now be manufactured quite cheaply.

parchment A sheet of material obtained from the skin of goats, sheep, or other animals. Vegetable parchment is a counterpart, made from chemical wood pulp.

plastic Capable of being molded or modelled, as in thermoplastic.

plastics A group of materials of high molecular weight which, while solid in the finished state, can be made to flow by the application of temperature or pressure.

plasticiser An agent or compound which, when added to plastics, lacquers or papers, for example, impart either softness or flexibility. Some of those in the manufacture of PVC cling films are alleged to be carcinogenic to certain laboratory animals, but any effect on humans has yet to be proven.

ply A fold or layer, as in a lamination.

polyethylene (PE) A synthetic material with excellent resistance to acids and alkalis and has no known solvent at room temperatures. It can be purchased in a great many weights and densities, each with different characteristics. Seek advice in its specification and use.

polyester The base material used for making polyethylene terephthalate (PET) and food trays for microwave products. It is also used as a substrate for audio and videocassette tape. In packaging its main use is in metallized materials such as tinsel.

polyethylene terephthalate (PET) This almost unbreakable plastics material is being used for an increasing number of containers - from carbonated beverages to jams, instant coffee and spirits.

polystyrene (PS) A thermoplastic material that is tasteless and odourless, with excellent water and weather resistance. It can be used for most foods and a range of other materials, with the exception of essential oils, petroleum products and turpentine which can deform the material.

polyvinyl chloride (PVC) A wide ranging packaging material, PVC can be used in its rigid form as bottles for squashes and cordials, while its plasticized form can be used in the generation of cling-type films and lidding materials for modified atmosphere packaging among other things.

polyvinylidene chloride (PVdC) This material is tough, transparent and resists the passage of a great many chemicals and gases. Du Pont's proprietary version is called Saran and the film is often used as a coating, of PET or glass bottles for example.

polypropylene (PP) This material was chiefly used as a rigid material for components, but has since become a major medium for the designer of biscuit and confectionery packaging. Oriented polypropylene has largely replaced paper wrappers in these two areas.

primary package A container such as a can, bottle, or jar, which directly holds the product.

reel Any device on which a material may be wound, hence the phrase 'from the reel'.

regenerated cellulose This term refers to films made from a cellulose base.

register To have one part positioned correctly with respect to another. Most often applied in printing.

sheet film (qv) thicker than 0.01in (0.025cm).

shelf life The expected timespan a product - often food - remains saleable. The current trend is to declare shorter and shorter shelf lives, so that foodstuffs are consumed before potentially harmful bacteria from the air can have any deleterious effects.

shoulder That part of a container between the main body and the neck, as on a glass bottle or collapsible tube.

shrinkwrap A wrap of flexible film, PE or PVC usually, on an object or collection of objects to hold them in place - a piggyback promotion offer for instance, or a tray of 20 cans of beer.

silk screen printing Ink is forced onto a container through a design on a taut silk screen. It is usually used for printing ceramic labels on glass containers, or for printing on polyethylene containers.

solid fibreboard Heavy solid board, commonly available at 0.06in (0.015cm), 0.08in (0.02cm), 0.10in (0.025cm), 0.12in (0.03cm) and 0.14in (0.035cm) thick. Made of two liners and a filler of chipboard. Used largely in shipping containers, but also in spirally wound tubes and packs for powdered detergents.

tear tape A strong tape, glued to the inside horizontal circumference of cassette tapes, biscuits, pharmaceutical packs and corrugated cartons, with one end protruding. Pulling the tape rips open the container.

tinplate Sheet steel of a special formula and temper, coated on both sides with a layer of pure tin. Its use as a container has revived in recent years, spearheading the nostalgia trend in container packaging.

underprint To print in register before the final design is printed.

vacuum packaging Packaging from which almost all the air has been removed prior to the final sealing of the container. The pack - whether rigid or flexible - must be constructed from barrier materials that retain the vacuum. The process usually extends the expected life of the product by protecting it from gases or water vapour in the atmosphere.

web A roll of paper, film or foil as it moves through a processing machine.

Index

(Figures in italics refer to captions)